he Elements

Gases

VIII		IB	IIB	IIIA	IVA	VA	VIA	VIIA	0
$s^2)^{\dagger}$	$(d^8s^2)^{\dagger}$	s^1d^{10}	s^2	s^2p^1	s^2p^2	s^2p^3	s^2p^4	s^2p^5	s^2p^6
									2 He 4.003
				5 B 10.81	6 C 12.01	7 N 14.01	8 O 15.999	9 F 18.99	10 Ne 20.18
				13 Al 26.98	14 Si 28.09	15 P 30.97	16 S 32.06	17 Cl 35.45	18 Ar 39.95
27 Co .93	28 Ni 58.71	29 Cu 63.54	30 Zn 65.37	31 Ga 69.72	32 Ge 72.59	33 As 74.92	34 Se 78.96	35 Br 79.91	36 Kr 83.80
45 Rh)2.91	46 Pd 106.4	47 Ag 107.87	48 Cd 112.40	49 In 114.82	50 Sn 118.69	51 Sb 121.75	52 Te 127.60	53 I 126.90	54 Xe 131.30
77 Ir 92.2	78 Pt 195.09	79 Au 196.97	80 Hg 200.59	81 Tl 204.37	82 Pb 207.19	83 Bi 208.98	84 Po	85 At	86 Rn

nner Transition Elements - *f*

62 Sm 0.35	63 Eu 151.96	64 Gd 157.25	65 Tb 158.92	66 Dy 162.50	67 Ho 164.93	68 Er 167.26	69 Tm 168.93	70 Yb 173.04	71 Lu 174.97
94 Pu	95 Am	96 Cm	97 Bk	98 Cf	99 Es	100 Fm	101 Md	102 No	103 Lw

PERGAMON INTERNATIONAL LIBRARY
of Science, Technology, Engineering and Social Studies

The 1000-volume original paperback library in aid of education, industrial training and the enjoyment of leisure

Publisher: Robert Maxwell, M.C.

The Chemistry of
VANADIUM, NIOBIUM AND TANTALUM

Comprehensive Inorganic Chemistry

The Chemistry of

VANADIUM

R. J. H. Clark

NIOBIUM AND TANTALUM

D. Brown

Chapters 34 and 35 of
Comprehensive Inorganic Chemistry

PERGAMON PRESS

OXFORD . NEW YORK . TORONTO
SYDNEY . PARIS . BRAUNSCHWEIG

Pergamon Press Offices:

U.K.	Pergamon Press Ltd., Headington Hill Hall, Oxford, OX3 0BW, England
U.S.A.	Pergamon Press Inc., Maxwell House, Fairview Park, Elmsford, New York 10523, U.S.A.
CANADA	Pergamon of Canada Ltd., 207 Queen's Quay West, Toronto 1, Canada
AUSTRALIA	Pergamon Press (Aust.) Pty. Ltd., 19a Boundary Street, Rushcutters Bay, N.S.W. 2011, Australia
FRANCE	Pergamon Press SARL, 24 rue des Ecoles, 75240 Paris, Cedex 05, France
WEST GERMANY	Pergamon Press GmbH, D-3300 Braunschweig, Postfach 2923, Burgplatz 1, West Germany

First edition 1973

Reprinted, with corrections, from Comprehensive Inorganic Chemistry, 1975

Library of Congress Catalog Card No. 77-189736

Printed in Great Britain by A. Wheaton & Co, Exeter

ISBN 0 08 018866 4 (hard cover)
ISBN 0 08 018865 6 (Flexicover)

CONTENTS

PREFACE

The excellent reception that has been accorded to *Comprehensive Inorganic Chemistry* since the simultaneous publication of the five volumes of the complete work has been accompanied by the plea that sections should be made available in a form that would enable specialists to purchase copies for their own use. To meet this demand the publishers have decided to issue selected chapters and groups of chapters as separate editions. These chapters will, apart from the corrections of misprints and the addition of prefatory material and individual indices, appear just as they did in the main work. Extensive revision would delay publication and greatly raise the cost, so limiting the circulation of these definitive reviews.

A. F. TROTMAN-DICKENSON
Executive Editor

34. VANADIUM

R. J. H. CLARK
University College London

1. THE ELEMENT

1.1. DISCOVERY AND HISTORY[1–3]

The discovery of element 23 was first claimed in 1801 by Andres Manuel del Rio, a professor of mineralogy in the University of Mexico. He recognized the existence of the new element in a lead ore from Zimapan in Mexico, and considered that it was similar to chromium and uranium in chemical properties. Owing to its formation of a variety of colours in acid media, del Rio gave the new element the name erythronium. Subsequently, however, del Rio retracted the claim, considering after further examination that the mineral was a basic chromate of lead. The retraction was substantiated by Collet-Descotils.

The element was not unequivocally discovered until 1830 and then by the Swede Sefström in an iron ore from Taberg, Sweden. In collaboration with Berzelius, the oxide of the new element was isolated. Sefström gave the element the name vanadium, after Vanadis, the legendary Norse goddess of beauty, in recognition of the richness of the colours of its derivatives. Wöhler in 1831 identified vanadium with the original erythronium of del Rio. Indeed, it is now known that the original ore from Zimapan was vanadinite.

The greatest contributor to the early chemistry of vanadium was H. E. Roscoe, who, in a series of papers in the period 1868–71, laid the foundation for most of the subsequent work. Working with copper ores mined in Cheshire, he prepared V_2O_5, VO_2, V_2O_3, VO, $VOCl_3$, $VOCl_2$, VOCl and other compounds and carefully studied their properties.

1.2. OCCURRENCE AND DISTRIBUTION

Vanadium has an abundance in the earth's crust of about 0·02%. It is widely spread—there being over 65 minerals known—but there are few concentrated deposits[4]. Small amounts of vanadium have been found associated with deposits of lead, lead–copper and lead–zinc.

The most important minerals are patronite, the polysulphide VS_4, vanadinite $3Pb_3(VO_4)_2.PbCl_2$ (which has the same structure as fluorapatite and chlorapatite), carnotite $K(UO_2)(VO_4)1{\cdot}5H_2O$, roscoelite $2K_2O.2Al_2O_3.(Mg, Fe)O.3V_2O_5.10SiO_2.4H_2O$ (a vanadium-bearing mica), mottramite $(Pb, Ca, Cu)_3(VO_4)_2$, descloizite $(Pb, Zn)_3(VO_4)_2$ and tyuya-

1 M. E. Weeks, *J. Chem. Ed.* **9** (1932) 863.
2 S. G. Sjöberg, *J. Chem. Ed.* **28** (1951) 294.
3 M. S. Vallarta and A. A. Y. Freg, *Nature*, **160** (1947) 163.
4 W. Rostoker, *The Metallurgy of Vanadium*, John Wiley and Sons, New York (1958).

munite, a uranium- and calcium-bearing mica, $Ca(UO_2.VO_4)_2.4H_2O$. It also occurs in some crude oils in the form of organic complexes.

The United States is the largest producer and consumer of vanadium, the bulk of the output being a by-product of the accelerated uranium production programme[5] in operation between 1952 and 1956. At the present time it continues to be a coproduct of the processing of the carnotite ores of Colorado, which contain 1–2% vanadium. The only other leading producers of vanadium in the Western World are Finland and South Africa, although several other potentially recoverable deposits are known to exist, for example, in Northern Rhodesia, Peru, Venezuela and France.

The element has also been identified in the sun, in stars, in meteorites and in a variety of plant and animal life in which it appears to play an important rôle.

1.3. METALLURGICAL ASPECTS

Most methods for the production of the metal involve prior roasting of the vanadium-containing ore to the oxide or to the metavanadate[4]. Vanadium is most commonly prepared in the form of ferrovanadium, a material which may contain 35–95% of vanadium. The oxide may be reduced by carbon, silicon or aluminium. Carbon is the cheapest reducing agent, but it is difficult to keep the carbon content of the product low.

Although aluminium is a much more expensive reducing agent, the resulting product is relatively pure and has a high vanadium content. Perhaps the most widely used reducing agent is ferrosilicon, the reaction being carried out in an air-tight electric furnace. Lime converts the silica produced to calcium silicate.

$$2V_2O_5+5Si \rightarrow 4V+5SiO_2$$
$$SiO_2+CaO \rightarrow CaSiO_3$$

The pure metal is very difficult to prepare, principally because it combines readily with oxygen, carbon, hydrogen and nitrogen and because it forms solid solutions with some of the products. Three principal general methods for the production of the pure metal have been used[6]: (i) reduction of chlorides, (ii) reduction of oxides and (iii) thermal dissociation of the diiodide. These are outlined separately.

(i) Reduction of vanadium tetrachloride may be accomplished using sodium, hydrogen over magnesium at 700°C, sodium hydride or apparently best (99·95% pure) by hydrogen in argon at 680–700°C. Vanadium oxytrichloride impurity in the tetrachloride is to be avoided because, subsequent to the reduction step, the oxygen appears as an interstitial element in the metal lattice, with deleterious effects on the ductility of the metal.

The separation of VCl_4 from $VOCl_3$ is difficult and cannot be accomplished satisfactorily by fractional distillation. This problem may be circumvented by effecting instead the reduction of the trichloride, e.g. with hydrogen at 900°C, or with magnesium turnings at 750–800°C in an argon atmosphere. The dichloride is also reduced to the metal by these reducing agents. Any hydrogen associated with the metal may be removed by heating *in vacuo* to ~ 1000°C.

(ii) Reduction of the pentoxide with a suitable reducing agent, such as mischmetall or (better) calcium, is also an effective synthetic method. The best procedure appears to be to

[5] E. H. Crabtree and F. L. Smith, *J. Less-Common Metals*, 3 (1961) 433.
[6] G. Brauer, *Handbook of Preparative Inorganic Chemistry*, 2nd ed., Academic Press, New York (1965).

reduce the pentoxide at 900–950°C in an iron tube with calcium, anhydrous calcium chloride and a trace of sodium or potassium being used to flux the CaO to produce a fluid slag at the temperatures generated within the bomb[7]. Calcium iodide may replace calcium chloride for this purpose [8, 9]. Alternative methods are also available[10] involving the analogous reduction of the sesquioxide V_2O_3.

(iii) The third general method for the production of the metal, due to van Arkel[11], involves the thermal dissociation of the diiodide on an incandescent tungsten filament. The impure metal and iodine are sealed *in vacuo* into a silica tube containing the filament. On heating the tube to 900–1000°C, the diiodide is produced; this has a significant vapour pressure, and is transported to and decomposed in the immediate vicinity of the filament, the metal being deposited on it with a purity $> 99{\cdot}5\%$. Any oxygen or nitrogen remain behind as oxides or nitrides and the iodine is recirculated. Owing to the comparatively low volatility of the diiodide, the method is not as good in the case of vanadium as for the Group IV elements. The method has been refined by Nash *et al.*[12].

Finally, it is also possible to produce very high purity ($\sim 99{\cdot}99\%$) metal by electro-refining calcium-reduced vanadium in either bromide or chloride electrolytes[13], e.g. in a medium of (in weight %), 51 KCl, 41 LiCl and 8 VCl_2. Reduction of vanadium dichloride to the metal occurs at the cathode, on which the vanadium is deposited. Simultaneously at the anode, oxidation and dissolution of the vanadium from the anode material takes place. The success of the electrorefining process is dependent on the selective solution of vanadium into the electrolyte. The first cycle of refining consists of using calcium-reduced vanadium as the anode material. In the second cycle of refining, the products of the first cycle are used as the anode material. The refined vanadium is deposited on a molybdenum cathode rod in each case. The hardness of the final product is 48 on the Rockwell F scale.

1.4. NUCLEAR PROPERTIES

Vanadium has only two naturally occurring isotopes with mass numbers 50 and 51. The percentage isotopic abundances of the isotopes are given in Table 1, together with the atomic masses of the individual nuclides (^{12}C scale). Also in Table 1 are the directly measured nuclear spins of the nuclides, their half-lives and some details of the modes of decay, the radiations emitted and their energies (in MeV). Where different modes of decay are possible, these are listed in order of decreasing abundance if this information is available. The thermal neutron cross-section for the naturally occurring mixture of isotopes is 5·00 barn[14].

Vanadium is the most stable nuclide with the mass number 51. Its stability relative to ^{51}Ti, ^{51}Cr and ^{51}Mn is shown in Fig. 1, which also includes details of the principal decay modes for the unstable isobars.

7 J. W. Marden and M. Rich, *Ind. Eng. Chem.* **19** (1927) 786.
8 R. K. McKechnie and A. U. Seybolt, *J. Electrochem. Soc.* **97** (1950) 311.
9 A. P. Beard and D. D. Crooks, *J. Electrochem. Soc.* **101** (1954) 597; *ibid.* **102** (1955) 361.
10 E. D. Gregory, W. C. Lilliendahl, and D. M. Wroughton, *J. Electrochem. Soc.* **98** (1951) 395.
11 A. E. van Arkel, *Metallwirtschaft* **13** (1934) 405; *Reine Metalle*, Springer, Berlin (1939).
12 V. W. Nash, H. R. Ogden, R. E. Durtschi and I. E. Campbell, *J. Electrochem. Soc.* **100** (1953) 272.
13 T. A. Sullivan, *J. Metals*, **17** (1965) 45; K. P. V. Lei and T. A. Sullivan, *J. Less-Common Metals*, **14** (1968) 145.
14 G. Friedlander, J. W. Kennedy and J. M. Miller, *Nuclear and Radiochemistry*, 2nd ed., John Wiley and Sons, New York (1964).

TABLE 1. NUCLIDES OF VANADIUM[a]

Nuclide	%	Isotopic mass	Nuclear spin	Half-life	Decay modes, radiations, energies	Cross-sections for thermal neutrons (barns)
^{46}V	—	45·96023	—	0·4s	β^+ 6·05	—
^{47}V	—	46·95489	—	31m	β^+ 1·89; γ	—
^{48}V	—	47·95226	—	16·1d	β^+ 0·69; EC; γ	—
^{49}V	—	48·94852	7/2	330d	EC	—
^{50}V	0·24	49·947165	6	6×10^{15}y	β^- 0·4; EC; γ	300
^{51}V	99·76	50·943978	7/2	—	—	4·5
^{52}V	—	51·94480	—	3·77m	β^- 2·6; γ	—
^{53}V	—	52·94337	—	2m	β^- 2·5; γ	—
^{54}V	—	—	—	55s	β^- 3·3; γ	—

EC = electron capture; β-particle energies in MeV.

[a] G. Friedlander, J. W. Kennedy and J. M. Miller, *Nuclear and Radiochemistry*, 2nd ed., John Wiley and Sons, New York (1964).

TABLE 2. PROPERTIES OF THE GROUP VA ELEMENTS

Property	Vanadium	Niobium	Tantalum
Atomic weight	50·942	92·91	180·95
Electronic structure	$3d^3 4s^2$	$4d^4 5s^1$	$5d^3 6s^2$
Ground state[a]	$^4F_{3/2}$	$^6D_{1/2}$	$^4F_{3/2}$
Melting point (°C)	1919[b]	2468[c]	2996[c]
Heat of fusion, ΔH_m (kcal/m)	—	3·51[c]	6·3–7·5[c]
Boiling point (°C)	3400[d]	4927[c], 3300[d]	5425 ± 100[c]
Heat of vaporization, ΔH_v (kcal/m)	—	91·1[c]	—
Absolute entropy $S^\circ_{298\cdot15}$ (cal/deg/m)	7·01[d]	—	—
First ionization potential[a] I_1 (eV)	6·74	6·88	7·88
Second ionization potential[a] I_2 (eV)	14·65	14·32	—
Third ionization potential[a] I_3 (eV)	29·31	25·04	22·3
Fourth ionization potential[a] I_4 (eV)	48·0	38·3	33·1
Fifth ionization potential[a] I_5 (eV)	65·2	50·0	45
Density			
20°C, X-ray (g/cc)	6·092	8·581	16·677
20°C, pyk (g/cc)	6·1	8·57	16·6
Radius, 20°C (Å)	1·3112[d]	1·429[f]	1·430[f]
Lattice constant, body centred cubic, 20°C (Å)	3·0282[d]	3·3005[f]	3·3029[f]
Magnetic susceptibility[g], $10^6\chi_A$, 25°C (cgsu)	255	204	154
Thermal conductivity[c] (cal/cm²/cm/sec/°C)	0·074 (100°C)	0·130	0·130
Electrical resistivity[c], 20°C (micro-ohm-cm)	24·8–26·0	16·0	12·45 (25°C)
Thermal neutron absorption cross section[h] (barns)	5·00	1·16	
Heat of atomization, ΔH_f°, 298 (kcal/g atom)	122·5	172·5	186·8

[a] C. E. Moore, *Nat. Bur. Standards* (*U.S.*) *Circ.* No. 467, Vols. I (1949), II (1952) and III (1958) and appendices.

[b] *JANAF Thermochemical Tables*, U.S. Air Force Contract AF 04(611)–7554, Aug. 1965.

[c] *Metals Handbook*, 8th ed., Vol. 1, American Society for Metals, Ohio, U.S.A. (1961).

[d] P. Pascal, *Nouveau Traité de Chimie Minérale*, Tome XII, Masson, Paris (1958).

[e] S. Beatty, *J. Metals*, 4 (1952) 987.

[f] J. W. Edwards, R. Speiser and H. L. Johnson, *J. Appl. Phys.* **22** (1951) 424.

[g] G. Foex, *Constantes Sélectionnées Diamagnétism et Paramagnétisme*, Masson, Paris (1957).

[h] G. Friedlander, J. W. Kennedy and J. M. Miller, *Nuclear and Radiochemistry*, 2nd ed., John Wiley and Sons, New York (1964).

[i] V. I. Vedeneyev, L. V. Gurvich, V. N. Kondrat'yev, Y. A. Medvedev and Ye. L. Frankevich, *Bond Energies, Ionization Potentials and Electron Affinities*, Arnold, London (1966).

The atomic weight of vanadium[15] on the ^{16}O scale is 50·95 based on the ratio $VOCl_3/3Ag$. That on the ^{12}C scale becomes 50·942.

1.5. PHYSICAL, CHEMICAL AND BIOLOGICAL PROPERTIES

Vanadium is a greyish-silver metal which crystallizes, like niobium and tantalum, with a body-centred cubic lattice (space group O_h^9–$Im3m$, $Z = 2$) the parameters for which are given in Table 2, together with many other physical data on these elements. Vanadium is believed[16] to undergo a phase transition at 1550°C, to an allotrope of unknown structure, but the evidence for this is inconclusive. There is also some evidence[4] for a further allotrope

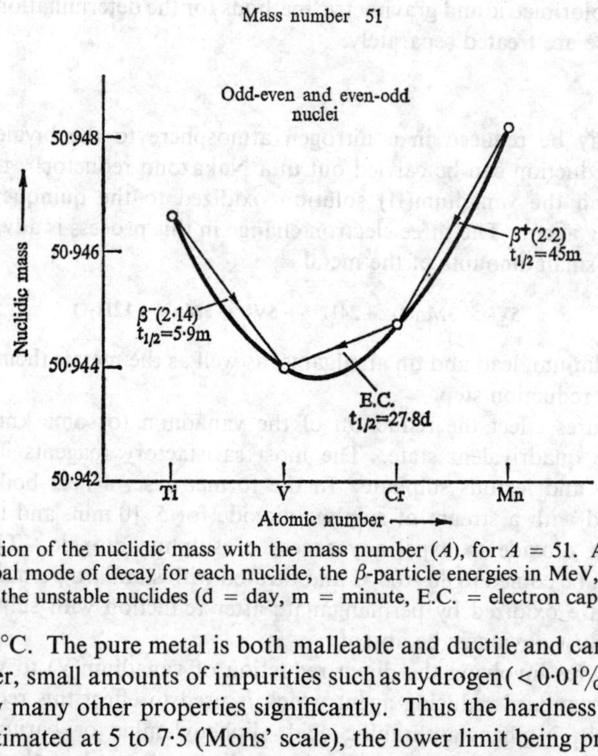

FIG. 1. Variation of the nuclidic mass with the mass number (A), for $A = 51$. Also indicated are the principal mode of decay for each nuclide, the β-particle energies in MeV, and the half-lives of the unstable nuclides (d = day, m = minute, E.C. = electron capture).

stable below −29°C. The pure metal is both malleable and ductile and can be cold-worked into foil. However, small amounts of impurities such as hydrogen (< 0·01%) cause embrittlement, and modify many other properties significantly. Thus the hardness of the metal has been variously estimated at 5 to 7·5 (Mohs' scale), the lower limit being probably the more nearly correct for the pure metal. The metal is weakly paramagnetic, has a thermal conductivity similar to that of iron, and has a density (~ 6·1) intermediate between that of titanium and iron. The metal has good structural strength and a low-fission neutron cross-section which gives it uses in nuclear technology. The metal is also used in producing rust-resistant spring and high speed tool steels. Other metallurgical properties are summarized by Rostocker[4].

Vanadium has good corrosion resistance to salt water, to air at ordinary temperatures, to sulphuric and hydrochloric acids and to alkalis[4]. However, it is attacked by concentrated sulphuric, nitric and hydrofluoric acids. At high temperatures it combines with most non-metals, e.g. with oxygen above 660°C it is oxidized via the lower oxides to V_2O_5. Many of

[15] A F. Scott and C. R. Johnson, *J. Am. Chem. Soc.* **52** (1930) 2638.
[16] A. U. Seybolt and H. T. Sumsion, *Trans. Am. Inst. Mining Met. Engrs.* **197** (1953) 292.

these compounds with non-metals are non-stoichiometric and interstitial, and some (e.g. with nitrogen and carbon) raise the melting point of the metal. The resistance of vanadium to corrosion at high temperatures by liquid metals such as sodium or bismuth–lead alloys is very good[4].

The element exists in trace quantities in much plant and animal life, but its precise function does not seem to have been elucidated. The subject has been reviewed in detail by Morette[17]. The maximum permitted quantity of V_2O_5 dust in the air is 0·5 mg/m^3 and of V_2O_5 fumes is 0·1 mg/m^3.

1.6. ANALYTICAL CHEMISTRY

Volumetric, colorimetric and gravimetric methods for the determination of vanadium are all in use[18]. These are treated separately.

Volumetric

Vanadium may be reduced in a nitrogen atmosphere to the bivalent state by zinc amalgam. The reduction can be carried out in a Nakazono reductor[19], the amalgam then being run off, and the vanadium(II) solution oxidized to the quinquevalent state with standard oxidizing agent. The three electron change in this process is advantageous for the determination of small amounts of the metal

$$5V^{2+}+3MnO_4^-+24H^+ \rightarrow 5V^{5+}+3Mn^{2+}+12H_2O$$

Alternatively, cadmium, lead and tin amalgams, as well as the metals themselves, have been used to effect the reduction step.

Other procedures effect the reduction of the vanadium to some known intermediate state, usually the quadrivalent state. The most satisfactory reagents in this regard are sulphur dioxide[20] and ferrous sulphate. In the former case, a near-boiling vanadium(V) solution is treated with a stream of sulphur dioxide for 5–10 min, and then with a rapid stream of carbon dioxide to expel any excess of sulphur dioxide. The vanadium(IV) solution may then be cooled to 80–100°C and titrated with standard permanganate solution. Elements which are oxidized by permanganate after reduction with sulphur dioxide (e.g. Fe, As and Sb) must obviously be absent.

Many procedures involving the direct reduction of vanadium(V) to vanadium(IV) are also in use; one standard reducing agent which is used to effect this reduction is ferrous sulphate, ferroin[21], N-phenylanthranilic acid, diphenylamine or barium diphenylamine sulphate being used as an indicator. The last indicator is used in the determination of vanadium in steels[18].

Colorimetric

Hydrogen peroxide is the most commonly used reagent for the colorimetric determination of vanadium[22]. When hydrogen peroxide is added to a solution containing up to 0·1 mg

17 A. Morette, in *Nouveau Traité de Chimie Minérale*, ed. P. Pascal, Masson, Paris (1958).
18 C. L. Wilson and D. W. Wilson, *Comprehensive Analytical Chemistry*, Vol. 1C, Elsevier, Amsterdam (1962).
19 T. Nakazono, *Sci. Rept. Tohoku Univ.* **14** (1925) 109; *J. Chem. Soc. Japan*, **42** (1921) 526, 761.
20 M. C. Steele and F. M. Hall, *Anal. Chim. Acta*, **9** (1953) 384.
21 G. H. Walden, L. P. Hammett and S. M. Edmonds, *J. Am. Chem. Soc.* **56** (1934) 57.
22 E. R. Wright and M. G. Mellon, *Ind. Eng. Chem., Anal. Ed.* **9** (1937) 375.

vanadium per ml, in sulphuric acid solution, a reddish-brown peroxo complex is formed. The optical density of such a solution is measured against a reagent blank having the same acidity and containing the same volume of hydrogen peroxide ($\sim$ 0·03%) at 450 or 290 mμ[23]. A large excess of hydrogen peroxide should be avoided, as this reduces the intensity of the colour, and also changes the colour to yellow. Elements which form coloured complexes with hydrogen peroxide, e.g. Ti, Cr, Mo(VI) and W, obviously interfere. Interference from titanium may be overcome by adding fluoride ion or hydrofluoric acid; these reagents at the same time remove any yellow colour caused by the presence of iron(III). If no titanium is present, any iron(III) may be decolorized using phosphate ion. Interference due to tungsten may be eliminated with oxalic acid.

An alternative and more sensitive colorimetric method involves the use of phosphotungstovanadic acid[23]. A yellow solution is formed on adding phosphoric acid ($\sim$ 0·5 M) and sodium tungstate ($\sim$ 0·025 M) to an acid vanadate solution, the optical density of which is recorded at 400 mμ and compared with that of known standards at this wavelength.

A colorimetric method based on the reaction between vanadium(V) and 8-hydroxyquinoline is also very sensitive[24].

Gravimetric

In general the gravimetric methods for the determination of vanadium are less specific than the volumetric and colorimetric methods, although they may be satisfactory in the absence of interfering elements. The various procedures have been assessed by Morette and Duval[25, 26].

One gravimetric method involves ignition of the vanadium-containing material to the pentoxide. This method is not particularly satisfactory, as the oxide melts at 658°C and thus slowly volatilizes; the vapour is also rather poisonous. Moreover, in the presence of interfering elements, the vanadium must be first selectively precipitated as some suitable complex, and the latter ignited to the oxide. Recommended reagents[18] in this regard are mercurous nitrate and cupferron.

In the absence of other metals forming insoluble oxinates (e.g. Fe, Ti, Al, Zr, Mo, W, Co or Ni) vanadium may be determined gravimetrically as the oxine complex[27] $V_2O_3(C_9H_6ON)_4$.

2. COMPOUNDS

2.1. ALLOYS

Vanadium forms alloys with a large number of elements[28], e.g. Al, Sv, As, Be, B, C, Cr, Co, Cu, Ge, Au, Hf, H, Fe, Mn, Hg, Mo, Ni, Nb, N, O, Pd, P, Pt, Re, Rh, Ru, Se, Si, Ag, S, Ta, Te, Th, Sn, Ti, W, U and Zr. Vanadium has an atomic radius of 1·36 Å, which is not appreciably different from that of a large number of other elements; hence a variety of solid solutions with other elements is possible, particularly for those cases in which the other element is chemically similar to vanadium, e.g. chromium.

[23] I. M. Kolthoff and P. J. Elving, *Treatise on Analytical Chemistry*, Part II, Vol. 8, Interscience, New York (1963).
[24] N. A. Talvitie, *Anal. Chem.* **25** (1953) 604.
[25] A. Morette, *Bull. soc. chim. France*, **17** (1950) 526.
[26] C. Duval and A. Morette, *Compt. rend.* **230** (1950) 545.
[27] R. Niericker and W. D. Treadwell, *Helv. Chim. Acta*, **29** (1946) 1472.
[28] W. Rostocker, *The Metallurgy of Vanadium*, John Wiley and Sons, New York (1958).

Those alloys for which there appears to be evidence for definite compound formation are as follows: V_5Al_8, VAl_3, VAl_6, VAl_{11}, VSb_2, V_3As, VAs, VBe_2, VB, V_3B_4, VB_2, V_2C, VC, V_3Co, VCo_3, V_3Ge, V_2Hf, $VH_{0\cdot94}$, V_3Ni, VNi_2, VNi_3, VN, VO, V_2O_3, V_3O_5, V_6O_{13}, V_2O_5, VP, V_3Pt, V_3Rh, VSe, V_2Se_3, VSe_2, V_3Si, V_5Si_3, VSi_2, VS, V_2S_3, VTe, V_3Sn, V_2Zr. The lattice parameters, structural types and other properties of these alloys have been given,[28] but their detailed properties are beyond the scope of this review. Many ternary alloy systems involving vanadium are also known[28].

The interstitial alloys formed with H, O, N, P, As, C and Si have been reviewed in detail separately[29], and are discussed later under the appropriate headings.

2.2. HYDRIDE

Vanadium in a hydrogen atmosphere at high temperatures can absorb considerable quantities of hydrogen. Thus, iodide vanadium on treatment with hydrogen at 1 atm pressure and 300°C for 5 hr forms a hydride with the final composition $VH_{0\cdot94}$ (48·4 atm.% of hydrogen)[30]. X-ray diffraction work has shown that the powder product is a single phase, with a tetragonal unit cell and lattice parameters $a = 3{\cdot}02$, $c = 3{\cdot}36$ Å. The hydride is a grey, metallic material which increases in brittleness with increasing hydrogen content. The magnetic susceptibility of the metal is decreased by the presence of hydrogen[31].

2.3. HALIDES AND OXYHALIDES

The halides, which are very important as starting materials for syntheses of new compounds, vary markedly in properties from the strongly reducing bivalent halides to the oxidizing pentafluoride. Their general properties[32] and preparations[32, 33] have been reviewed. Monohalides, VX, are unknown; where X = Cl, Br and I, the heats of formation are calculated to be positive[34]. For the fluoride, the heat of formation at room temperature is calculated to be -18 kcal/m; however, the compound would be thermodynamically unstable with respect to disproportionation as follows[34]:

$$2VF \rightarrow VF_2 + V \qquad \Delta H = -120 \text{ kcal/m.}$$

Dihalides

Preparation. The difluoride was prepared for the first time in 1966 by Stout and Boo[35] by reduction of the trifluoride at 1150°C with a controlled mixture of HF and H_2. The compound was obtained as blue needles possessing the rutile structure. It has a peak in its heat capacity at 7·0°K presumably associated with long range magnetic ordering (Tables 3, 4).

The dichloride may be prepared by heating the trichloride for 1 hr at 430°C in a current of hydrogen[36] or by disproportionation of the trichloride at ~800°C in a stream of

29 H. J. Goldschmidt, *Interstitial Alloys*, Butterworths, London (1967).
30 W. Rostoker, *The Metallurgy of Vanadium*, John Wiley and Sons, New York (1958).
31 T. R. P. Gibb and W. A. Norder, *Inorg. Chem.* **5** (1966) 1947.
32 R. J. H. Clark, *The Chemistry of Titanium and Vanadium*, Elsevier, Amsterdam (1968).
33 G. Brauer, *Handbook of Preparative Inorganic Chemistry*, 2nd ed., Academic Press, New York (1965).
34 M. Barber, J. W. Linnett and N. H. Taylor, *J. Chem. Soc.* (1961) 3323.
35 J. W. Stout and W. O. J. Boo, *J. Appl. Phys.* **37** (1966) 966.
36 W. Klemm and E. Hoschek, *Z. anorg. und allgem. Chem.* **226** (1936) 359.

TABLE 3. PROPERTIES OF THE HALIDES OF VANADIUM(II)

Halide	M.W.	Colour	Subl. temp. °C	ρ^{20} (pyk) g/cc	$\Delta H_{f,298}$ kcal/m	S°_{298} cal/deg/m	μ B.M.
VF_2	88·94	blue	—	3·96[a]	—	—	3·36 (300°K)[b]
VCl_2	121·85	pale green	~910[c]	3·09[d]	−110[e]	23·2[f]	2·41 (300°K)[g]
VBr_2	210·76	brown-orange	~800[h]	4·52 (25°C)[i]	−83[h]	29[f]	2·80 (300°K)[g]
VI_2	304·74	red	750–800 vac.[j]	5·0 (0°C)[i]	−63[k]	32[f]	3·21 (288°K)[l]

[a] J. W. Stout and W. O. J. Boo, *J. Appl. Phys.* **37** (1966) 966 (X-ray density).
[b] J. W. Stout and H. Y. Lau, *J. Appl. Phys.* **38** (1967) 1472.
[c] M. A. Oranskaya and I. L. Perfilova, *Russ. J. Inorg. Chem.* (Eng. Transl.) **6** (1961) 131.
[d] W. Klemm and L. Grimm, *Z. anorg. und allgem. Chem.* **249** (1942) 198; the densities of VBr_2 and VI_2 were determined by X-ray means.
[e] S. A. Shchukarev, M. A. Oranskaya, T. A. Tolmacheva and Y. S. Il'inskii, *Russ. J. Inorg. Chem.* (Eng. Transl.) **5** (1960) 3.
[f] P. G. Nelson and A. G. Sharpe, *J. Chem. Soc.(A)* (1966) 501.
[g] W. Klemm and E. Hoschek, *Z. anorg. und allgem. Chem.* **226** (1936) 361.
[h] S. A. Shchukarev, T. A. Tolmacheva and V. M. Tsintsius, *Russ. J. Inorg. Chem.* (Eng. Transl.) **7** (1962) 345.
[i] F. Ephraim and E. Ammann, *Helv. Chim. Acta*, **16** (1933) 1273.
[j] A. Morette, *Compt. rend.* **207** (1938) 1218; V. M. Tsintsius and E. E. Yudovich, *Russ. J. Inorg. Chem.* (Eng. Transl.) **9** (1964) 555.
[k] T. A. Tolmacheva, V. M. Tsintsius and L. V. Andrianova, *Russ. J. Inorg. Chem.* (Eng. Transl.) **8** (1963) 281.
[l] W. Klemm and L. Grimm, *Z. anorg. und allgem. Chem.* **249** (1942) 209.

TABLE 4. CRYSTALLOGRAPHIC DATA ON HALIDES AND OXYHALIDES

Compound	Space group	Lattice parameters (Å)	ρ (g/cc) X-ray	Z	Ref.
VF_4	Hexagonal	$a = 5·38$, $c = 5·16$	3·25	2	a
$VOCl_2$	Orthorhombic	$a = 3·842$, $b = 11·761$, $c = 3·383$	2·99	2	b
VF_3	Rhombohedral $R\bar{3}c$–D_{3d}^6	$a = 5·373$, $\alpha = 57°31'$	3·46	2	c
VCl_3	Hexagonal $R\bar{3}$–C_{3i}^2	$a = 6·045$, $c = 17·45$	2·87	6	d, e
VBr_3	Hexagonal $R\bar{3}$–C_{3i}^2	$a = 6·400$, $c = 18·53$	4·41	6	e
VI_3	Hexagonal $R\bar{3}$–C_{3i}^2	$a = 6·919$, $c = 19·91$	5·20	6	k, l
VOCl	Orthorhombic $Pmmn$–D_{2h}^{13}	$a = 3·78$, $b = 3·30$, $c = 7·91$	3·45	2	f
VOBr	Orthorhombic $Pmmn$–D_{2h}^{13}	$a = 3·775$, $b = 3·380$, $c = 8·425$	4·53	2	j
VF_2	Tetragonal $P4_2/mnm$–D_{4h}^{14}	$a = 4·80$, $c = 3·24$	3·96	2	g
VCl_2	Hexagonal $P\bar{3}m1$–D_{3d}^3	$a = 3·60$, $c = 5·83$	3·09	1	h
VBr_2	Hexagonal $P\bar{3}m1$–D_{3d}^3	$a = 3·768$, $c = 6·180$	4·61	1	i
VI_2	Hexagonal $P\bar{3}m1$–D_{3d}^3	$a = 4·058$, $c = 6·753$	5·25	1	i, l

[a] R. G. Cavell and H. C. Clark, *J. Chem. Soc.* (1962) 2692.
[b] R. J. Sime, *Z. Krist.* **124** (1967) 3.
[c] K. H. Jack and V. Gutmann, *Acta Cryst.* **4** (1951) 246.
[d] W. Klemm and E. Krose, *Z. anorg. Chem.* **253** (1947) 218.
[e] R. E. McCarley, J. W. Roddy and K. O. Berry, *Inorg. Chem.* **3** (1964) 50.
[f] H. Schäfer and F. Wartenpfuhl, *J. Less-Common Metals*, **3** (1961) 29.
[g] J. W. Stout and W. O. J. Boo, *J. Appl. Phys.* **37** (1966) 966.
[h] P. Ehrlich and H. J. Seifert, *Z. anorg. und allgem. Chem.* **301** (1959) 282.
[i] W. Klemm and L. Grimm, *Z. anorg. und allgem. Chem.* **249** (1942) 198.
[j] C. Levayer and J. Rouxel, *Compt. rend.* **268C** (1969) 167.
[k] K. O. Berry, R. R. Smardzewski and R. E. McCarley, *Inorg. Chem.* **8** (1969) 1994.
[l] D. Juza, D. Giegling and H. Schäfer, *Z. anorg. und allgem. Chem.* **366** (1969) 121.

nitrogen[37]. In the latter method, the tetrachloride also produced is carried away by the nitrogen stream. The dichloride may also be prepared by treating vanadium metal at 1000°C with hydrogen chloride.

The dibromide may likewise be prepared by reduction of the tribromide with a stream of hydrogen. However, the diiodide is prepared[38] by thermal decomposition of the triiodide above ~ 280°C. The iodine produced in the reaction may be removed by pumping *in vacuo*. Alternatively, the diiodide may be produced[39] by direct reaction of a stoichiometric mixture of the elements in a short quartz tube *in vacuo* at 160–170°C.

Properties. The dihalides are coloured paramagnetic solids (Table 3) which, apart from the difluoride, all possess the cadmium iodide structure (the lattice parameters are given in Table 4). On the bases of temperature range studies down to 90°K, it is thought that the dichloride, dibromide and diiodide are antiferromagnetic with Néel points of approximately 700, 400 and 100°K respectively. That of the difluoride is 42°K.

The diffuse reflectance spectrum of the dichloride consists[40] of three bands, at 9000, 14,000 and 21,500 cm^{-1}, which may be assigned to the three spin-allowed *d–d* transitions expected for a d^3 ion in an octahedral ligand field, i.e. $^4T_{2g} \leftarrow {}^4A_{2g}$, $^4T_{1g}(F) \leftarrow {}^4A_{2g}$ and $^4T_{1g}(P) \leftarrow {}^4A_{2g}$. These data are consistent with the following values for the two ligand field parameters: $Dq = 900$ cm^{-1}, $B = 606$ cm^{-1}.

The dihalides are all hygroscopic, and strongly reducing; hence they must be handled in the absence of air. In water, they all form the $V(H_2O)_6^{2+}$ ion. Little research into their coordination chemistry has yet been carried out, but it appears from research on the dichloride, that complexes with nitrogen- and oxygen-donor ligands are formed with some facility.

Expressions for the vapour pressures of VCl_2 and VBr_2 are given in Table 5.

Trihalides

Preparation. The trifluoride may be prepared by passing anhydrous gaseous HF in a stream of nitrogen over the trichloride or tribromide held at ~ 600°C in a platinum crucible[41]. It may also be prepared[42] by heating the dichloride in gaseous HF at 700°C, by heating $VH_{0.72}$ with HF at 400°C, by disproportionation[43] of the tetrafluoride at 100–120°C and, together with other fluorides, by direct fluorination of the metal[42] at 200°C.

The trichloride is prepared[44] by thermal decomposition of the tetrachloride on reflux for ~ 2 days at ~ 150°C. Any remaining tetrachloride together with any oxytrichloride may be distilled off at 200°C in a stream of hydrogen. Alternatively, the trichloride can be formed[33, 45] by heating the sesquioxide V_2O_3 with thionyl chloride in a bomb for 24 hr at 200°C.

$$V_2O_3 + 3SOCl_2 \rightarrow 2VCl_3 + 3SO_2$$

The bomb is then cooled to 0°C and opened to allow the sulphur dioxide to escape. The trichloride is washed with carbon disulphide and dried *in vacuo* at 80°C. A further

37 P. Ehrich and H. J. Seifert, *Z. anorg. und allgem. Chem.* **301** (1959) 282.
38 A. Morette, *Compt. rend.* **207** (1938) 1218.
39 W. Klemm and L. Grimm, *Z. anorg. und allgem. Chem.* **249** (1942) 198.
40 R. J. H. Clark, *J. Chem. Soc.* (1964) 417.
41 O. Ruff and H. Lickfett, *Ber.* **44** (1911).
42 H. J. Emeléus and V. Gutmann, *J. Chem. Soc.* (1949) 2979.
43 R. G. Cavell and H. C. Clark, *J. Chem. Soc.* (1962) 2692.
44 F. Ephraim and E. Ammann, *Helv. Chim. Acta*, **16** (1933) 1273.
45 H. Hecht, G. Jander and H. Scheapmann, *Z. anorg. Chem.* **254** (1947) 255.

method[46] involves reflux of the pentoxide with sulphur monochloride for eight hours. The excess of the latter, containing dissolved sulphur, is then decanted off, and the resulting trichloride is washed with carbon disulphide and dried in vacuum at ~ 120°C.

$$2V_2O_5+6S_2Cl_2 \rightarrow 4VCl_3+5SO_2+7S$$

The trichloride may also be prepared by refluxing the powdered metal with iodine monochloride. After cooling the mixture, the iodine is removed in carbon tetrachloride and the trichloride dried *in vacuo*[47].

Vanadium tribromide may be prepared by the action of dry bromine on the following substances[32, 33]; vanadium nitride at red heat; the carbide at 500–600°C; the metal in a

TABLE 5. VAPOUR PRESSURES OF HALIDES AND OXYHALIDES $\log_{10}P(\text{at}) = A - B/T$

Halide	Process	A	B	Range (°C)	ΔH kcal/m	ΔS cal/deg/m	Ref.
VF_5	s → v	8·168	2,608	−20 to 19·5	11·94	37·4	a
VF_5	l → v	7·549	2,423	19·5 to 45·5	10·64	34·5	a
$VOCl_3$	l → v	4·82	1,923	19 to 98	8·79	22·0	b
VCl_4	l → v	4·700	1,998	39·5 to 80·2	9·14	21·5	c
	l → v	5·21	2,185	24·8 to 85·4	10·0	23·8	d
VCl_3	s → v	8·32	9,777	352 to 467	44·7	38·0	c
VBr_3	s → v	8·24	9,470	317 to 427	43·3	37·7	e
VCl_2	s → v	5·833	9,804	490 to 631	44·8	26·6	e
	s → v	5·725	9,720·7	910 to 1100	44·4	26·2	f
VBr_2	s → v	5·9	9,830	800 to 905	44·9	27·0	g
	s → v	6·20	10,460	514 to 716	47·8	28·3	e

[a] H. C. Clark and H. J. Emeléus, *J. Chem. Soc.* (1957) 2119.
[b] A. V. Komandin and M. L. Vlodavets, *Zh. Fiz. Khim.* **26** (1952) 1291.
[c] J. H. Simons and M. G. Powell, *J. Am. Chem. Soc.* **67** (1945) 75.
[d] S. A. Shchukarev, M. A. Oranskaya, T. A. Tolmacheva and A. K. Yakhkind, *Russ. J. Inorg. Chem.* (Eng. Transl.) **1** (1956) 30.
[e] R. E. McCarley and J. W. Roddy, *Inorg. Chem.* **3** (1964) 60.
[f] M. A. Oranskaya and I. L. Perfilova, *Russ. J. Inorg. Chem.* (Eng. Transl.) **6** (1961) 131.
[g] V. M. Tsintsius and E. E. Yudovich, *Russ. J. Inorg. Chem.* (Eng. Transl.) **9** (1964) 555.

sealed, evacuated tube at 400°C; ferrovanadium at red heat, the bromine being carried in a current of dry carbon dioxide[44]; an intimate mixture of the pentoxide and carbon. It may also be prepared by the action of IBr on the powdered metal[47], or by heating the pentoxide with carbon tetrabromide in a Pyrex ampoule[48] for 10–12 hr at 350°C.

The triiodide is best prepared by a modification[49] of Morette's original method[38]. Sheet vanadium metal is heated with iodine in an evacuated silica tube at ~ 300°C for 1–2 days. The triiodide which forms on the surface of the sheet is easily removed, and any excess of iodine may be sublimed from the product by heating *in vacuo* at 80–100°C.

Properties. The trihalides are coloured, crystalline solids, all of which are polymeric

[46] H. Funk and W. Weiss, *Z. anorg. und allgem. Chem.* **295** (1958) 327.
[47] V. Gutmann, *Monatsh. Chem.* **81** (1950) 1155.
[48] S. A. Shchukarev, T. A. Tolmacheva and V. M. Tsintsius, *Russ. J. Inorg. Chem.* (Eng. Transl.) **7** (1962) 777.
[49] T. A. Tolmacheva, V. M. Tsintsius and L. V. Andrianova, *Russ. J. Inorg. Chem.* (Eng. Transl.) **8** (1963) 281.

involving six-coordinate vanadium atoms. In the trifluoride, the fluorine atoms are not close packed although their arrangement approximates to hexagonal close packing[50]. The V–F distance is 1·95 Å. The trichloride and tribromide have hexagonal or rhombohedral layer lattices of the bismuth triiodide type (Table 4); the V–Cl distance in the trichloride is 2·45 Å.

The magnetic moments of the trihalides (Table 6) are all slightly lower than the spin-only value for two unpaired electrons (2·83 B.M.). The trichloride and tribromide are considered to be antiferromagnetic, with θ values of 35° and 20° respectively over the range 90–293°K.

TABLE 6. PROPERTIES OF THE HALIDES AND OXYHALIDES OF VANADIUM(III)

Compound	M.W.	Colour	Dec. temp. (°C)	ρ_{20}(pyk) g/cc	$\Delta H_f°$, 298 kcal/m	$S°_{298}$ cal/deg/m	μ B.M.
VF_3 (s)	107·94	green	1406 [a] [b]	3·36 (19°C) [b]	−319±20 [c]	—	2·55 (293°K) [d]
VCl_3 (s)	157·30	red-violet	425 disprop. [e]	2·82 [f]	−134 [g]	31·3 [h]	2·74 (293°K) [i]
VBr_3 (s)	290·67	grey-brown	400 disprop. [j]	4·20 (25°C) [q]	−118 [k]	40 [h]	2·67 (293°K) [i]
VI_3 (s)	431·64	brown-black	280 dec. vac. [l]	5·14 [m]	−67 [l]	44 [h]	2·92 (295°K) [m]
VOCl (s)	102·39	brown	620 dec. vac. [n]	3·44 (25°C) [n]	−144 [o]	18 [o]	0·78 (295°K) [n]
VOBr (s)	146·85	violet	~480 dec. [p]	4·00 (18°C) [p]	—	—	—

[a] B. J. Sturm and C. W. Sheridan, *Inorg. Synth.* **7** (1963) 87; no decomposition in this case.
[b] O. Ruff and H. Lickfett, *Ber.* **44** (1911) 2539.
[c] R. G. Cavell and H. C. Clark, *J. Chem. Soc.* (1965) 444; the value is a calculated, not an experimental one.
[d] R. S. Nyholm and A. G. Sharpe, *J. Chem. Soc.* (1952) 3579.
[e] M. A. Oranskaya, Yu. S. Lebedev and I. L. Perfilova, *Russ. J. Inorg. Chem.* (Eng. Transl.) **6** (1961) 132.
[f] W. Klemm and E. Krose, *Z. anorg. Chem.* **253** (1947) 218.
[g] P. Gross and C. Hayman, *Trans. Faraday Soc.* **60** (1964) 45.
[h] P. G. Nelson and A. G. Sharpe, *J. Chem. Soc.* (*A*) (1966) 501.
[i] W. Klemm and E. Hoschek, *Z. anorg. und allgem. Chem.* **226** (1936) 359.
[j] R. E. McCarley and J. W. Roddy, *Inorg. Chem.* **3** (1964) 60.
[k] S. A. Shchukarev, T. A. Tolmacheva and V. M. Tsintsius, *Russ. J. Inorg. Chem.* (Eng. Transl.) **7** (1962) 345.
[l] T. A. Tolmacheva, V. M. Tsintsius and L. V. Andrianova, *Russ. J. Inorg. Chem.* (Eng. Transl.) **8** (1963) 281.
[m] D. Juza, D. Giegling and H. Schäfer, *Z. anorg. und allgem. Chem.* **366** (1969) 121.
[n] H. Schäfer and F. Wartenpfuhl, *J. Less-Common Metals*, **3** (1961) 29.
[o] H. Oppermann, *Z. anorg. und allgem. Chem.* **351** (1967) 127.
[p] O. Ruff and H. Lickfett, *Ber.* **44** (1911) 2534.
[q] F. Ephraim and E. Ammann, *Helv. Chim. Acta*, **16** (1933) 1273.

However, there is no evidence for an antiferromagnetic transition in the trifluoride down to 4·2°K. The diffuse reflectance spectra of the trihalides have been reported[40].

Vanadium trifluoride is insoluble in common organic solvents and almost insoluble in water[41]. The other trihalides, however, are very hygroscopic, forming the $V(H_2O)_6^{3+}$ ion in aqueous solution; this ion is easily hydrolysed, forming slightly acid solutions, and easily oxidized to the oxovanadium(IV) ion. With the possible exception of the trifluoride, all the trihalides are readily oxidized in air. The trichloride and tribromide form a range of complexes with oxygen-, nitrogen-, sulphur- and phosphorus-donor ligands in which the coordination number of the vanadium is five or six[32].

The trihalides behave differently on heating; whereas the trifluoride may be sublimed, the trichloride and tribromide partly disproportionate (to VX_2+VX_4), and partly sublime

[50] K. H. Jack and V. Gutmann, *Acta Cryst.* **4** (1951) 246.

(Table 5), and the triiodide decomposes (to VI_2+I_2). In the last case, the dissociation has been followed between 300 and 530°C with the result that the iodine pressure may be represented as follows[49].

$$\log_{10}P(\text{at}) = 5{\cdot}9 - 4800/T$$

There is no significant region of homogeneity between the di- and triiodides. Vanadium trichloride and vanadium tribromide form a continuous solid-solution phase over the complete composition range[51].

Oxyhalides

Only two oxyhalides, the oxychloride and oxybromide, have been characterized to date. The oxychloride is prepared[52] by a chemical transport reaction between the sesquioxide V_2O_3 and the trichloride using a temperature gradient of 620–720°C. It may also be prepared by heating the oxydichloride in a stream of nitrogen[37]. It is a brown, slightly paramagnetic material which belongs to the orthorhombic FeOCl structural type (Table 4). The oxybromide is similarly prepared either by thermal decomposition of the oxydibromide at ~ 360°C *in vacuo*, or by a sealed tube reaction between VBr_3 and As_2O_3 held at 400°C for 6 days.

The chemistry of these compounds has been little investigated.

Tetrahalides

Preparation. Vanadium tetrafluoride was first made in 1911 by Ruff and Lickfett[41] by treating vanadium tetrachloride with anhydrous HF at −28°C. A purer and more homogeneous product is obtained[43] if the synthesis is carried out in an inert solvent such as trichlorofluoromethane, which remains liquid at −78°C. The tetrafluoride is also obtained along with the tri- and penta-fluorides, by direct fluorination of the metal; it can be separated therefrom owing to the different volatilities of the three fluorides.

The tetrachloride may be prepared by the action of dry chlorine on vanadium metal[53] at 250–300°C, on the carbide[54], nitride, or silicide of vanadium, or on ferrovanadium. It may also be synthesized by disproportionation of the trichloride at 425°C or above[54], or by the action of various chlorinating agents (e.g. $SOCl_2$, SO_2Cl_2 and $COCl_2$) on the metal or on the pentoxide[55] at 350°C. It is purified by fractional distillation in a chlorine atmosphere and freed from dissolved chlorine by repeated freezing and pumping *in vacuo*.

The tetrabromide was first isolated in 1964 by McCarley and Roddy[56]. Their method involves disproportionation of the tribromide at 325°C, and trapping the resulting tetrabromide vapour on a condenser held at −78°C.

The tetraiodide is unknown.

Properties. Many physical data on the tetrahalides are given in Tables 4, 5 and 7. The tetrafluoride is structurally different from the other two halides. Its low solubility in non-polar solvents and low volatility suggest that it is a fluorine-bridged polymer of VF_6 units, each unit sharing four fluorine atoms with adjacent units. Its exact structure is, however, not known.

[51] R. E. McCarley, J. W. Roddy and K. O. Berry, *Inorg. Chem.* 3 (1964) 50.
[52] H. Schäfer and F. Wartenpfuhl, *J. Less-Common Metals*, 3 (1961) 29.
[53] P. Gross and C. Hayman, *Trans. Faraday Soc.* **60** (1964) 45.
[54] O. Ruff and H. Lickfett, *Ber.* **44** (1911) 506.
[55] E. Chauvenet, *Compt. rend.* **152** (1911) 87.
[56] R. E. McCarley and J. W. Roddy, *Inorg. Chem.* 3 (1964) 54.

TABLE 7. PROPERTIES OF THE HALIDES AND OXYHALIDES OF VANADIUM(IV)

Compound	M.W.	Colour	M.p.°C	B.p.°C	ρ_{20} g/cc	$\Delta H_f^\circ{}_{,298}$ kcal/m	S°_{298} cal/deg/m	μ B.M.
VF_4 (s)	126·93	green	100 subl. & disprop.[a]	—	3·15[a]	−321[b]	—	1·68(293°K)[c]
VCl_4 (l)	192·75	dark-brown	−25·7[d]	152[d]	1·820(25·3°C)[e]	−136·2[f]	58[g]	1·61(300°K)[i]
VCl_4 (g)							87·6[h]	
VBr_4 (g)	370·58	magenta	−23 dec.[j]	—	—	−108·5[k]	~80[k]	—
VOF_2 (s)	104·94	yellow	—	—	3·396(19°C)[l]	—	—	—
$VOCl_2$ (s)	137·85	green	~300 disprop.[g]	—	2·88(13°C)[m]	−165[g]	28·5[g]	—
$VOBr_2$ (s)	226·76	yellow-brown	~320 dec.[n]	—	—	—	—	—

[a] R. G. Cavell and H. C. Clark, *J. Chem. Soc.* (1962) 2692.
[b] R. G. Cavell and H. C. Clark, *Trans. Faraday Soc.* **59** (1963) 2706.
[c] R. J. H. Clark, *J. Chem. Soc.* (1963) 1377.
[d] J. H. Simons and M. G. Powell, *J. Am. Chem. Soc.* **67** (1945) 75.
[e] A. G. Whittaker and D. M. Yost, *J. Chem. Phys.* **17** (1949) 188.
[f] P. Gross and C. Hayman, *Trans. Faraday Soc.* **60** (1964) 45.
[g] H. Oppermann, *Z. anorg. und allgem. Chem.* **351** (1967) 113.
[h] J. A. Creighton, J. H. S. Green and W. Kynaston, *J. Chem. Soc.* (*A*) (1966) 208.
[i] R. J. H. Clark and D. J. Machin, *J. Chem. Soc.* (1963) 4430.
[j] R. E. McCarley and J. W. Roddy, *Inorg. Chem.* **3** (1964) 54.
[k] S. A. Shchukarev, T. A. Tolmacheva and V. M. Tsintsius, *Russ. J. Inorg. Chem.* (Eng. Transl.) **7** (1962) 345.
[l] O. Ruff and H. Lickfett, *Ber.* **44** (1911) 2539.
[m] H. E. Roscoe, *J. Chem. Soc.* **21** (1868) 322.
[n] O. Ruff and H. Lickfett, *Ber.* **44** (1911) 2534.

The tetrachloride and tetrabromide, on the other hand, are tetrahedral monomeric species in so far as is known. Precise electron diffraction work on the tetrachloride indicates[57] that it has T_d symmetry and V–Cl bond lengths of 2·138±0·002 Å; electronic spectral and magnetic measurements indicate that this structure is preserved in the liquid and solid states[58]. It probably crystallizes in a molecular lattice. It is non-conductor of electricity. Little is known about the properties of the magenta tetrabromide, owing to the fact that it decomposes at −23°C into the tribromide and bromine.

Vanadium tetrafluoride is a lime-green, paramagnetic powder which hydrolyses to a brown powder and finally to a blue paste on exposure to moisture. The brown product reported by Ruff and Lickfett[41] was presumably surface-hydrolysed. The tetrafluoride disproportionates at 100–120°C *in vacuo* to VF_3 and VF_5, but significant disproportionation occurs even at room temperature if the compound is sealed in a dry glass tube at atmospheric pressure for several days. However, the tetrafluoride also partly sublimes at 100–120°C.

The tetrafluoride reacts with either bromine trifluoride or fluorine to give the pentafluoride, and with liquid nitryl fluoride in the presence of iodine pentafluoride to form[43] the complex $NO_2^+VF_6^-$. It is a weaker fluorinating agent then the pentafluoride. It forms[43] 1:1 adducts with pyridine, ammonia and SeF_4 which may be fluorine-bridged polymers; its Lewis acid capacity has otherwise been scarcely investigated.

Vanadium tetrachloride is a red-brown liquid which hydrolyses extremely rapidly in moist air to produce in water a solution of the oxovanadium(IV) ion. It is soluble in non-polar solvents without change, but with potential ligands it reacts rapidly and violently to form 1:1 or 1:2 addition compounds; these are similar to those formed by titanium tetrachloride. The chlorine is partly or completely replaced in hydroxylic solvents.

The tetrachloride is only just thermodynamically stable[59]. Even when in a Pyrex tube over a pressure of chlorine, it slowly decomposes into the trichloride and chlorine. The decomposition may be catalytically or photochemically induced. The heat and entropy of melting at 252·7°K are 2·3±1·5 kcal/m and 9·1±0·6 cal/deg/m respectively[60].

In the ground state the tetrachloride has the e^1 configuration. Its magnetic moment (1·61 B.M., independent of temperature)[58] and ligand field spectrum ($^2T_2 \leftarrow {}^2E =$ 7900 cm^{-1} = $10Dq$) (Fig. 2) are consistent with this. However, this band has three components, at 6600, 7880 and 9010 cm^{-1} the existence of which is thought[61] to be a consequence of the Jahn–Teller effect (which requires a nuclear configurational instability for the undistorted molecule). The charge transfer spectra of the molecule have also been assigned according to the molecular orbital scheme developed for other tetrahedral molecules[62] (Table 8).

The vibrational spectrum of the tetrachloride is very similar to that of titanium tetrachloride. In terms of T_d nomenclature, the four fundamentals are assigned as follows[63]: $\nu_1(a_1)$ 383, $\nu_2(e)$ 105±20, $\nu_3(t_2)$ 487, $\nu_4(t_2)$ 129 cm^{-1}. The doubly degenerate bending mode (e) has not been observed directly. The dielectric constant[64] of the tetrachloride is 3·107±0·007 at 25°C, and the dipole moment is near zero.

57 Y. Morino and H. Uehara, *J. Chem. Phys.* **45** (1966) 4543.
58 R. J. H. Clark and D. J. Machin, *J. Chem. Soc.* (1963) 4430.
59 J. H. Simons and M. G. Powell, *J. Am. Chem. Soc.* **67** (1945) 75.
60 H. Oppermann, *Z. phys. Chem.* (*Leipzig*), **236** (1967) 161.
61 C. J. Ballhausen and J. de Heer, *J. Chem. Phys.* **43** (1965) 4304.
62 A. Carrington and D. S. Schonland, *Mol. Phys.* **3** (1960) 331.
63 J. A. Creighton, J. H. S. Green and W. A. Kynaston, *J. Chem. Soc.* (*A*) (1966) 208.
64 C. N. Caughlan and F. Cartan, *J. Am. Chem. Soc.* **81** (1959) 3840.

Mixed chloride–bromides VBr_xCl_{4-x} ($x = 1$, 2 or 3) are believed to exist in the gas phase on passing bromine over heated vanadium trichloride, but they have not been isolated. Two brown chloro-fluorides have been prepared[65] by adding AsF_3 to VCl_4 in the required quantities, but their structures are unknown; they are $VClF_3$ and $V_2Cl_3F_5$. The compound VCl_2F_2 has also been reported.

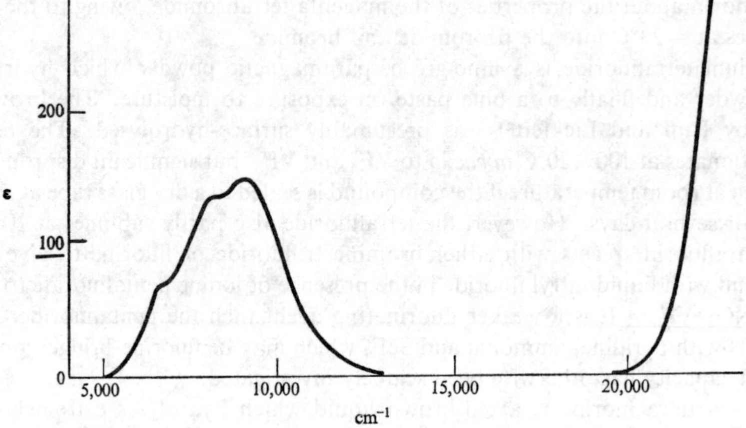

FIG. 2. Ligand field spectrum of vanadium tetrachloride.

Oxydihalides

The oxydihalides of vanadium(IV) have not yet been investigated in much detail.

Preparation. The oxydifluoride is prepared[41] by heating the oxydibromide in a stream of anhydrous HF, initially at 150–200°C, and then for 6 hr at 600–700°C.

The oxydichloride was originally prepared by Roscoe by reduction of the oxytrichloride with zinc powder at 400°C in a sealed tube. A more efficient method[33, 46] involves the reaction between the pentoxide, the trichloride and the oxytrichloride as follows:

$$V_2O_5+3VCl_3+VOCl_3 \rightarrow 6VOCl_2$$

The reactants are sealed in a tube and held in a temperature gradient, the hot end being at 600°C. The oxydichloride is transported out of the reaction zone in 4–5 days.

The oxydibromide may be prepared either by thermal decomposition of the oxytribromide at 180°C, or by the action of bromine vapour on a mixture of the pentoxide and sulphur held in a sealed tube at 500–600°C. In the latter case, the oxydibromide sublimes to the cooler end of the tube, this being held at 260°C. It can also be prepared by bromination of the sesquioxide.

The oxydiiodide has apparently not yet been prepared anhydrous.

Properties. Few physical properties and none of the structures of these compounds have been established (Table 7). The oxydifluoride is insoluble in most solvents, but it forms a hydrate, as do the other oxydihalides.

[65] L. Kolditz, V. Neumann and G. Kilch, *Z. anorg. und allgem. Chem.* **325** (1963) 275.

Pentafluoride

The only established pentahalide of vanadium is the pentafluoride.

Preparation. The pentafluoride may be prepared by disproportionation of the tetrafluoride[43] at 100–120°C, or by passing gaseous fluorine diluted with nitrogen (or bromine trifluoride) over the powdered metal contained in a nickel boat at 300°C[42]. The product is collected in a rigorously dried, all glass apparatus at −78°C and purified by trap-to-trap distillation from dry sodium fluoride. Apart from moisture, it is necessary to eliminate all traces of vacuum grease and hydrogen fluoride from the apparatus.

TABLE 8. ELECTRONIC SPECTRUM OF VCl_4 IN THE VAPOUR STATE[a]

Band maximum cm^{-1}	ϵ_{max}	Assignment
~7900	146	$(2e)^1 \rightarrow (3t_2)^1$
24,800	2800	$(t_1)^6(2e)^1 \rightarrow (t_1)^5(2e)^2$
33,900	4500	$(t_1)^6(2e)^1 \rightarrow (t_1)^5(3t_2)^1(2e)^1$
45,400 sh	~5000	$(2t_2)^6(2e)^1 \rightarrow (2t_2)^5(2e)^2$
> 50,000	> 12,000	$(2t_2)^6(2e)^1 \rightarrow (2t_2)^5(3t_2)^1(2e)^1$

[a] F. A. Blankenship and R. L. Belford, *J. Chem. Phys.* **36** (1962) 633; R. J. H. Clark and D. J. Machin, *J. Chem. Soc.* (1963) 4430; F. Pennella and W. J. Taylor, *J. Mol. Spect.* **11** (1963) 321.

TABLE 9. PROPERTIES OF THE HALIDES AND OXYHALIDES OF VANADIUM(V)

Compound	M.W.	Colour	M.p.°C	B.p.°C	ρ_{20} g/cc	$\Delta H_f°$, 298 kcal/m
VF_5 (s)	145·93	white	19·5[a]	48·3[a]	> 2·5[b]	−352[c]
VF_5 (l)					2·508[b]	
VOF_3 (s)	123·94	pale yellow	110 subl.[d]	—	2·459 (20·5)[b]	—
$VOCl_3$ (l)	173·30	yellow	−78·9[e]	127·2[e]	1·830[f]	−177[g]
$VOBr_3$ (l)	306·67	deep red	−59[h]	133 (100 mm)[i] 180 dec.[i]	2·993 (15°C)[j]	—
VO_2F (s)	101·94	brown	> 300[k]			—
VO_2Cl (s)	118·39	orange		180 dec.[l]	2·29[m]	−183[l]

[a] H. C. Clark and H. J. Emeléus, *J. Chem. Soc.* (1957) 2119.
[b] R. G. Cavell and H. C. Clark, *J. Chem. Soc.* (1963) 4261.
[c] R. G. Cavell and H. C. Clark, *Trans. Faraday Soc.* **59** (1963) 2706.
[d] L. E. Trevorrow, *J. Phys. Chem.* **62** (1958) 362.
[e] A. V. Komandin and M. L. Vlodavets, *Zh. Fiz. Khim.* **26** (1952) 1291.
[f] Estimated from the weighted mean of a number of measurements at different temperatures; see P. Pascal, *Nouveau Traité de Chimie Minérale*, Tome XII, Masson, Paris (1958).
[g] I. L. Perfilova, I. V. Kozlova, S. A. Shchukarev and I. V. Vasil'kova, *Vestn. Leningr. Univ., Ser. Fiz. i Khim.* **16** (1961) 125.
[h] A. G. Loomis and H. Schlundt, *J. Phys. Chem.* **19** (1915) 734.
[i] H. E. Roscoe, *Phil. Trans. Roy. Soc. London*, **160** (1870) 317.
[j] F. Gonzâléz Nuñez and E. Figueroa, *Compt. rend.* **206** (1938) 437.
[k] J. Weidlein and K. Dehnicke, *Z. anorg. und allgem. Chem.* **348** (1966) 278.
[l] H. Oppermann, *Z. anorg. und allgem. Chem.* **351** (1967) 113.
[m] K. Dehnicke, *Chem. Ber.* **97** (1964) 3354.

Properties. The pentafluoride is a white solid which melts[66] to a pale yellow viscous

[66] H. C. Clark and H. J. Emeléus, *J. Chem. Soc.* (1957) 2119; A. D. Beveridge and H. C. Clark, *Halogen Chemistry*, **3** (1967) 179.

liquid at 19·5°C. The viscosity of the liquid (1·24 poise at 25°C) together with its high Trouton constant (33·1) indicate that the compound is appreciably associated in the liquid state. The Raman spectrum of the liquid confirms this indication[67]. It is also thought that in this state, the pentafluoride partly ionizes to VF_4^+ and VF_6^- ions because of the relatively high value found for its specific conductivity[66] ($2·4 \times 10^{-4}$ cm^{-1} ohm^{-1} at 25°C).

In the solid state, the pentafluoride is undoubtedly also associated, although there are no structural data available. In the vapour state, however, the compound is monomeric with D_{3h} symmetry as evidenced by an analysis of its infrared and Raman spectrum. The following assignments of the fundamentals have been made[67]: 719 and 608 (a'_1), 784 and 331(a_2''), 810, 282 and 200 (e') and 350 (e''). Approximate bond stretching force constants on the modified Urey-Bradley force field are $K_{radial} = 5·51$ md/Å, $K_{axial} = 3·94$ md/Å. Slightly different force constants are given by Vizi *et al.*[68] who have also calculated the entropy for the ideal gaseous molecule at 1 atm pressure and 298·15°C ($S°_{298·15}$) to be 77·86 cal/deg/m. Electron diffraction data confirm that the gaseous molecule is trigonal bipyramidal[69] with equal V–F bond lengths of 1·71 Å.

Expressions for the vapour pressure of the pentafluoride are given in Table 5, and other properties are summarized in Table 9. The compound is believed to attack glass at room temperature, and is a powerful oxidizing and fluorinating agent.

Oxytrihalides

Preparation. The oxytrifluoride is prepared[41] by the action of oxygen on the trifluoride held at red heat in a platinum crucible, or by treatment of the pentoxide with fluorine[70] or bromine trifluoride[71].

The oxytrichloride may be prepared[41] by the action of dry chlorine on V_2O_3 or V_2O_5 at 600–800°C, but temperatures of 300–500°C are adequate in the presence of charcoal or sulphur. It may also be prepared by treating the pentoxide with hydrogen chloride, thionyl chloride or acetyl chloride, or by direct action of the trichloride with oxygen. The liquid is normally purified by fractional distillation in the presence of sodium.

The oxytribromide may be prepared by passing bromine vapour over V_2O_3 or over a mixture of V_2O_5 and carbon[72], temperatures of ~600°C being required to effect reaction. It is purified by distillation under reduced pressure.

The oxytriiodide is unknown.

Properties, The properties of the oxytrihalides are summarized in Table 9. They are all hygroscopic, the oxytrichloride and oxytribromide being violently so.

The oxytrifluoride attacks glass, especially when warm.

The compounds are all monomeric with C_{3v} symmetry in the vapour state, as evidenced by their vibrational spectra (Table 10). Electron diffraction data on the oxytrichloride vapour confirm this, the V–Cl and V–O bond lengths being 2·12 and 1·56 Å respectively[73], and the ClVCl angles being 111°. An expression for the vapour pressure of the oxytrichloride is

[67] H. H. Claassen and H. Selig, *J. Chem. Phys.* **44** (1966) 4039.
[68] B. Vizi, J. Brunvoll and A. Müller, *Acta Chem. Scand.* **22** (1968) 1279.
[69] G. V. Romanov and V. P. Spiridonov, *Zh. Strukt. Khim.* **7** (1966) 882.
[70] H. M. Haendler, S. F. Bartram, R. S. Becker, W. J. Bernard and S. W. Bukata, *J. Am. Chem. Soc.* **76** (1954) 2177.
[71] H. J. Emeléus and A. A. Woolf, *J. Chem. Soc.* (1950) 164.
[72] F. Gonzâléz Nunez and E. Figueroa, *Compt. rend.* **206** (1938) 437.
[73] K. J. Palmer, *J. Am. Chem. Soc.* **60** (1938) 2360.

given in Table 5; the vapour pressure amounts to 4·4 mm at 0°C. Other physical properties of the oxytrichloride are summarized in Table 11.

The oxytrichloride is completely miscible with many hydrocarbons and halides such as CCl_4, $TiCl_4$ and $SnCl_4$, and dissolves others such as PCl_3, $POCl_3$, $AsCl_3$ and $SbCl_3$. However, more polar solvents, and Lewis bases in general, react immediately with it to form addition

TABLE 10. VIBRATIONAL FREQUENCIES OF VANADIUM OXYTRIHALIDES IN THE VAPOUR STATE (cm^{-1})

	$\nu_1(a)$	$\nu_2(a)$	$\nu_3(a)$	$\nu_4(e)$	$\nu_5(e)$	$\nu_6(e)$
VOF_3[a]	1057·8	721·5	257·8	806	308	204·3
$VOCl_3$[b]	1035	408	165	504	249	129
$VOBr_3$[c]	1025	271	120	400	212	83

[a] H. Selig and H. H. Claassen, *J. Chem. Phys.* **44** (1966) 1404.
[b] F. A. Miller and L. R. Cousins, *J. Chem. Phys.* **26** (1957) 329.
[c] F. A. Miller and W. K. Baer, *Spectrochim. Acta*, **17** (1961) 112.

TABLE 11. OTHER PROPERTIES OF VANADIUM OXYTRICHLORIDE

Property	Value	Reference
Dipole moment	0·39 Debye	C. N. Caughlan and F. Cartan, *J. Am. Chem. Soc.* **81** (1959) 3840.
Viscosity, 20°C	0·007541 dyne sec/cm^2	G. P. Luchinskiĭ, *Zh. Obshch. Khim.* **7** (1937) 2116.
Heat and entropy of melting	2·29 kcal/m 11·7 cal/deg/m	H. Oppermann, *Z. phys. Chem.* (*Leipzig*), **32** (1967) 161.
Entropy, S°_{298}	82·2 cal/deg/m (Transl. 41·36, Rot. 26·65, Vib. 14·20)	K. K. Kelley and E. G. King, *U.S. Bur. Mines Bull.* No. 592 (1961) 98.
Refractive index, 27°C	1·6300 (λ = 5890Å)	C. N. Caughlan and F. Cartan, *J. Am. Chem. Soc.* **81** (1959) 3840.
Dielectric constant, 25°C[a]	2·898±0·007	C. N. Caughlan and F. Cartan, *J. Am. Chem. Soc.* **81** (1959) 3840.
Specific conductance, 20°C	9×10^{-12} $ohm^{-1}cm^{-1}$	V. Gutmann and S. Aftalion-Hinl, *Monatsh. Chem.* **84** (1953) 207.

[a] Dielectric constant of vanadium oxytribromide is 3·70 at 20°C, A. G. Loomis and H. Schlundt, *J. Phys. Chem.* **19** (1915) 734.

compounds which often have the stoichiometrics[74] $VOCl_3.L$ and $VOCl_3.2L$. Vanadium oxytribromide has similar chemical properties, but has been little investigated.

The charge transfer spectra of the oxytrichloride have been assigned on the basis of a molecular orbital treatment[75].

Dioxyhalides

The dioxyfluoride is prepared by fluorination of the dioxychloride[76]. The latter may be prepared[77] by passing a stream of Cl_2O gas diluted with oxygen into $VOCl_3$ at room

[74] H. L. Krauss and G. Gnatz, *Chem. Ber.* **95** (1962) 1023.
[75] F. A. Miller and W. B. White, *Spectrochim. Acta*, **9** (1957) 98; C. Dijkgraaf, *Spectrochim Acta*, **21** (1965) 969, 1419.
[76] J. Weidlein and K. Dehnicke, *Z. anorg. und allgem. Chem.* **348** (1966) 278.
[77] K. Dehnicke, *Chem. Ber.* **97** (1964) 3354.

temperature, by passing ozone into boiling $VOCl_3$ or by reaction between As_2O_3 and $VOCl_3$ as follows:

$$VOCl_3+Cl_2O \rightarrow VO_2Cl+2Cl_2$$
$$VOCl_3+O_3 \rightarrow VO_2Cl+Cl_2+O_2$$
$$3VOCl_3+As_2O_3 \rightarrow 3VO_2Cl+2AsCl_3$$

The dioxychloride is an orange, very hygroscopic crystalline solid which dissolves without apparent decomposition in solvents such as tetrahydrofuran or acetoacetic acid. It is probably polymeric in the solid state. It has some acceptor properties towards ligands such as pyridine.

2.4. OXIDES

A large number of different phases have been reported for the vanadium–oxygen system, but not all of these appear to be real. An incomplete phase diagram has been published by Rostocker[78], and a more complete one by Stringer[79], who has also reviewed the subject critically.

TABLE 12. VANADIUM OXIDE PHASES[a]

Oxide phase	Stability range	Structure	Lattice parameters
α	$VO_{0 \cdot 00}$ to $VO_{0 \cdot 01}$	b.c.c.	$a = 3{\cdot}025$ to $a = 3{\cdot}049$ Å
β	$VO_{0 \cdot 18}$ to $VO_{0 \cdot 33}$	b.c. tetragonal	$a = 2{\cdot}992$, $c = 3{\cdot}263$ to $a = 2{\cdot}943$, $c = 3{\cdot}523$
γ	$VO_{0 \cdot 53}$	hexagonal	—
δ	$VO_{0 \cdot 85}$ to $VO_{1 \cdot 24}$	NaCl	$a = 4{\cdot}024$ to $a = 4{\cdot}138$
ϵ	$VO_{1 \cdot 27}$	b.c. tetragonal	$a = 16{\cdot}623$, $c = 16{\cdot}515$
ζ	V_2O_3	α-Al_2O_3 (rhombohedral)	$a = 4{\cdot}952$, $c = 14{\cdot}002$
η	V_3O_5	monoclinic	$a = 9{\cdot}983$, $b = 5{\cdot}031$, $c = 9{\cdot}835$, $\beta = 138{\cdot}80°$
θ_n	V_nO_{2n-1} $4 < n < 8$	based on rutile	
e.g. θ_5	V_5O_9	triclinic	$a = 5{\cdot}475$, $b = 6{\cdot}994$, $c = 8{\cdot}718$, $\alpha = 97{\cdot}53°$, $\beta = 112{\cdot}44°$, $\gamma = 108{\cdot}99°$
θ	VO_2	monoclinic <70°C	$a = 5{\cdot}743$, $b = 4{\cdot}517$, $c = 5{\cdot}375$, $\beta = 122{\cdot}61°$
θ	VO_2	tetragonal rutile >70°C	$a = 4{\cdot}530$, $c = 2{\cdot}869$ (at 180°C)
K	V_6O_{13}	monoclinic	$a = 11{\cdot}90$, $b = 3{\cdot}671$, $c = 10{\cdot}122$, $\beta = 100{\cdot}87°$
λ	$VO_{2 \cdot 47}$ to $VO_{2 \cdot 5}$	orthorhombic	$a = 11{\cdot}519$, $b = 3{\cdot}564$, $c = 4{\cdot}373$

[a] J. Stringer, *J. Less-Common Metals*, **8** (1965) 1.

Primary Vanadium–Oxygen Solid Solution (α-phase)

Oxygen enters the body-centred cubic vanadium lattice interstitially up to the solubility limit of 3.2 at. % (1 wt. %)[80]. In so doing, it brings about a slight lattice expansion from 3·025 Å for the pure metal to 3·049 Å at the solubility limit, as well as a marked hardening of the metal from 125 to 450 VHN (Vickers hardness). A summary of the structures and lattice parameters of all the established oxide phases is given in Table 12.

78 W. Rostocker, *The Metallurgy of Vanadium*, Wiley, New York (1958).
79 J. Stringer, *J. Less-Common Metals*, **8** (1965) 1.
80 A. U. Seybolt and H. T. Sumsion, *Trans. A.I.M.E.* **197** (1953) 292.

The β-phase is stable over the range $VO_{0.18}$–$VO_{0.33}$, the Vickers hardness of which increases uniformly over the two-phase $\alpha+\beta$ region to a maximum of 1100 VHN at 15 at. % (5 wt. %) oxygen[82]. Little is known about the γ-phase.

Vanadium Monoxide (δ-phase)

The monoxide is formed as a grey, metallic powder by reduction of higher oxides with a number of different reducing agents, most usually with hydrogen. It has the sodium

TABLE 13. PROPERTIES OF THE PRINCIPAL VANADIUM OXIDES

Property	V_2O_5	V_2O_4	V_2O_3	VO
Colour	orange	blue	grey-black	grey
Melting point (°C)	658[a]	1637[b]	1967[b]	950[c] disprop. vac.
Density, pyk. (g/cc)	3·352[d] (22·6°C)	4·260[d] (21·4°C)	4·843[d] (22·0°C)	5·55[e] (25°C)
Heat of formation $\Delta H^\circ_{f,298\cdot15}$ (kcal/m)	−370·6[f]	−341·2[f]	−291·3[f]	−103·2[f]
Absolute entropy $S^\circ_{298\cdot15}$ (cal/deg/m)	31·3[d]	24·5[d]	23·5[d]	10·2[g]
Magnetic susceptibility $10^6\chi'_M$ (c.g.s.u.) (corrected for diamagnetism)	66[h] (293°K)	99[i] (293°K)	940[h] (293°K)	3376[f] (288°K)
Magnetic moment (B.M.)	~0	0·41[i]	1·49[h]	2·80[j]
Structure	distorted tetragonal pyramid[k]	α-form-distorted rutile[l]	corundum[m]	sodium chloride[n]
Space group	$Pmnm$–D_{2h}^{13}	$P2_1/c$–C_{2h}^5	$R\bar{3}c$–D_{3d}^6	$Fm3m$–O_h^5
Number of molecules/unit cell	2	4	2	4
V–O distances (Å)	1·585–2·02[k]	1·76–2·05[l]	1·96–2·06[m]	2·05[n]

[a] T. Carnelley, *J. Chem. Soc.* **33** (1878) 273.
[b] E. Friederich and L. Sittig, *Z. anorg. und allgem. Chem.* **145** (1925) 127.
[c] H. Hartmann and W. Mässing, *Z. anorg. Chem.* **266** (1951) 98.
[d] C. T. Anderson, *J. Am. Chem. Soc.* **58** (1936) 564.
[e] M. Frandsen, *J. Am. Chem. Soc.* **74** (1952) 5046.
[f] A. D. Mah and K. K. Kelley, *U.S. Bureau Mines Rept. Invest.* 5858 (1961).
[g] N. P. Allen, O. Kubaschewski and O. von Goldbeck, *J. Electrochem. Soc.* **98** (1951) 417.
[h] W. Klemm and E. Hoschek, *Z. anorg. und allgem. Chem.* **226** (1936) 359.
[i] W. Rüdorff, G. Walter and J. Stadler, *Z. anorg. und allgem. Chem.* **297** (1958) 1.
[j] E. Wedekind and C. Horst, *Ber.* **45** (1912) 262.
[k] H. G. Bachmann, F. R. Ahmed and W. H. Barnes, *Z. Krist.* **115** (1961) 110.
[l] A. Magnéli and G. Andersson, *Acta Chem. Scand.* **9** (1955) 1378; G. Andersson, *Acta Chem. Scand.* **10** (1956) 623.
[m] G. Andersson, *Acta Chem. Scand.* **8** (1954) 1599; R. E. Newnham and Y. M. de Haan, *Z. Krist.* **117** (1962) 235.
[n] N. Schönberg, *Acta Chem. Scand.* **8** (1954) 221.

chloride structure over the wide composition range[81] $VO_{0.80}$–$VO_{1.30}$ or[82] $VO_{0.85}$–$VO_{1.24}$. The X-ray density of stoichiometric VO (6·5 g/cc, based on the appropriate lattice parameter of 4·093 Å)[81] is some 15% higher than the pyknometric density (5·55 g/cc, Table 13). This indicates that the lattice is highly defective at the stoichiometric composition. At the vanadium-rich phase limit the cation lattice contains ~ 13% vacancies, but at the oxygen-rich phase limit the anion lattice is nearly perfect. It is thought that there is clustering of defects on the cation and anion lattices.

Other properties of the monoxide are summarized in Table 13. Like the higher oxides,

[81] N. Schönberg, *Acta Chem. Scand.* **8** (1954) 221; G. Andersson, *Acta Chem. Scand.* **8** (1954) 1599.
[82] P. V. Gel'd, S. I. Alyamovskii and I. I. Matveenko, *J. Struct. Chem. (U.S.S.R.)* **2** (1961) 286.

it is anti-ferromagnetic, and also shows a discontinuity in its electrical conductivity at $\sim 120°K$.

The ground term of the gaseous VO molecule is probably $^4\Sigma^-$; its vibrational frequency, equilibrium separation and dissociation energy are $1011{\cdot}6\ cm^{-1}$, 1·589 Å and 6·4 eV respectively[83].

A phase (ϵ) reported[84] to have a narrow range of homogeneity at $VO_{1{\cdot}27}$ has not been substantiated by subsequent work, and so its existence is in doubt.

Vanadium Sesquioxide (ζ-phase)

The sesquioxide, V_2O_3, may be prepared by reduction of the pentoxide with hydrogen, or with a number of other reducing agents (e.g. C or CO). It crystallizes with the corundum structure[85], the shortest V–V distances parallel and perpendicular to the hexagonal c-axis being 2·70 and 2·88 Å respectively[86]. The phase has only a narrow range of homogeneity.

The sesquioxide is antiferromagnetic with a transition temperature of 168°K. The transition is associated with a very large ($\sim 10^6$ times) change in the electrical resistivity of the compound, this change occurring within a single degree of temperature[87]. Both the anisotropy of the susceptibility of single crystals[88] of V_2O_3, as well as the magnetic properties[89] of the $V_2O_3/\alpha\text{-}Al_2O_3$ system have been studied. For other physical properties, see Table 13.

The optical spectrum of the compound contains two bands at $\sim$ 17,500 and $\sim 25{,}100\ cm^{-1}$ which have been assigned as the lowest two spin-allowed ligand field bands for the VO_6 chromophore, namely, the $^3T_{2g} \leftarrow {^3T_{1g}}(F)$ and the $^3T_{1g}(P) \leftarrow {^3T_{1g}}(F)$ transitions[90].

The Oxide V_3O_5

This phase, the η-phase, is reported to have a very narrow range of homogeneity[82].

The Homologous Series V_nO_{2n-1}

Five oxides have been identified[82] in the range V_3O_5 to VO_2; these have the compositions $VO_{1{\cdot}75}$, $VO_{1{\cdot}80}$, $VO_{1{\cdot}84}$, $VO_{1{\cdot}86}$ and $VO_{1{\cdot}87}$. They are evidently members of homologous series of general formula V_nO_{2n-1}, where $n = 4$, 5, 6, 7 or 8. The structures of all of these phases appear to be based on that of rutile, with periodic defects (Magnéli defects) resembling stacking faults which effectively introduce extra planes of metal atoms[79].

Vanadium Dioxide

The dioxide may be prepared by a mild reduction of the pentoxide, e.g. with V_2O_3, C, CO, SO_2 or oxalic acid. It forms deep blue crystals which, above 341°K, have the rutile structure. Below the transition temperature, however, the chains of octahedra are puckered and the metal atoms are shifted so as to form V–V pairs; alternate V–V distances along the

83 K. D. Carlson and C. Moser, *J. Chem. Phys.* **44** (1966) 3259.
84 S. Westman and C. Nordmark, *Acta Chem. Scand.* **14** (1960) 465.
85 R. E. Newnham and Y. M. de Haan, *Z. Krist.* **117** (1962) 235.
86 P. F. Bongers, *Solid State Commun.* **3** (1965) 275.
87 J. B. Goodenough, *Magnetism and the Chemical Bond*, Interscience, New York (1963).
88 P. H. Carr and S. Foner, *J. Appl. Phys.* **31** (1960) 344S.
89 A. Callaghan, M. J. Rossiter and F. S. Stone, *Trans. Faraday Soc.* **62** (1966) 3463.
90 D. S. McClure, *J. Chem. Phys.* **36** (1962) 2757.
91 A. Magnéli and G. Andersson, *Acta Chem. Scand.* **9** (1955) 1378; G. Andersson, *Acta Chem. Scand.* **10** (1956) 623.

chain are[81] 2·65 and 3·12 Å. The bonding within these chains is cooperative, the *d*-electron on each metal atom being trapped into localized V–V bonds so as to quench the metallic conductivity along the *c*-axis. In addition, the magnetic moment drops[92] from 1·41 to 0·53 B.M. over the range 352–337°K. Isomorphous substitution of V^{4+} by Ti^{4+} disturbs the anti-ferromagnetic interaction within the lattice leading to higher susceptibilities and higher conductances than for the pure dioxide.

As for VO and V_2O_3, the transformation temperature is lowered by raising the pressure[93].

The Phase V_6O_{13}

The existence of a phase between VO_2 and V_2O_5 was first suggested by Klemm and Hoschek[94] and then identified as V_6O_{13} by Andersson[82]. The stability of the phase is not, however, thought to be very great[95].

Vanadium Pentoxide

The pentoxide is the final product of the oxidation of vanadium metal, lower vanadium oxides, ammonium vanadate, or sulphides or nitrides of vanadium. It is brick-red, but it may appear blue-black if it has not been prepared in fully oxidizing conditions. The pentoxide may also be prepared by burning the finely divided metal in an excess of oxygen, or by thermal decomposition of ammonium metavanadate.

$$2NH_4VO_3 \rightarrow V_2O_5+2NH_3+H_2O$$

It is slightly soluble in water ($\sim$0·07 g/l) to form a pale yellow acidic solution which readily becomes colloidal; the properties of the vanadium sols so formed have been much studied. It dissolves in acids as well as in bases, and its moderate oxidizing properties are indicated by the fact that chlorine is evolved when it dissolves in hydrochloric acid; vanadium(IV) species are simultaneously produced.

From the definitive structural analysis of the compound[96], it is clear that each vanadium atom forms five bonds to oxygen atoms, one at 1·585 Å (O′), one at 1·80 Å (O‴), two at 1·88 Å (O″) and one at 2·02 Å (O″). The structurally different oxygen atoms O′, O‴ and O″ form one, two and three bonds, respectively, to vanadium atoms and so the coordination is best represented as $VO'(O''')\frac{1}{2}(3O'')\frac{1}{3}$. The angles between the VO′ and the other four VO bonds lie in the range 104·5–105°. There is also a sixth oxygen atom (from a different layer) in the position *trans* to the V–O′ bond and at the comparatively long distance of 2·785 Å. The compound has only a small range of homogeneity.

The pentoxide loses oxygen reversibly[97] in the region 700–1125°C, a phenomenon which may account for its important catalytic properties. The pentoxide is an important catalyst for many reactions. These include the following[98]: sulphur dioxide to the trioxide, the sulphonation of aromatic hydrocarbons or of pyridine, the reduction of olefins; the oxidation of hydriodic acid by bromic acid or by hydrogen peroxide, of sugar by nitric acid, of alcohol by air, of stannous salts by nitric or chloric acid, of cyclic organic compounds

92 M. S. Archer, D. S. P. Roebuck and F. J. Whitby, *Nature*, **174** (1954) 754.
93 S. Minomura and H. Nagasaki, *J. Phys. Soc. Japan*, **19** (1964) 131.
94 W. Klemm and E. Hoschek, *Z. anorg. und allgem. Chem.* **226** (1936) 359.
95 H. Flood and O. J. Kleppa, *J. Am. Chem. Soc.* **69** (1947) 998; N. P. Allen, O. Kubaschewski and O. von Goldbeck, *J. Electrochem. Soc.* **98** (1951) 417.
96 H. G. Bachmann, F. R. Ahmed and W. H. Barnes, *Z. Krist.* **115** (1961) 110.
97 E. F. Milan, *J. Phys. Chem.* **33** (1929) 498.
98 N. V. Sidgwick, *The Chemical Elements and Their Compounds*, Oxford University Press, Oxford (1950).

by hydrogen peroxide, of naphthalene and its substitution products by air, and the reduction of aromatic hydrocarbons by hydrogen.

The electrical conductivities of both single-crystal and molten V_2O_5 indicate that the compound is an oxygen-deficient semi-conductor which contains V^{4+} ions as defects[99].

2.5. OXOVANADIUM SPECIES

Oxovanadium(IV) Species

Many oxometal species have been characterized, the most stable of which are the UO_2^{2+} and VO^{2+} ions. Indeed, the latter is thought to be the most stable diatomic ion known. It forms a wide variety of stable complexes which may be cationic, neutral or anionic[100, 101].

Most oxovanadium(IV) complexes are of the types[102] $[VOL_5]^{n\pm}$, $[VOL_4]^{n\pm}$, $[VOL_xL'_{5-x}]^{n\pm}$ or $[VOL_xL'_{4-x}]^{n\pm}$ e.g. $R_3[VOF_5]$, $R_2[VOF_4]$, $VOCl_2.3DMSO$ (DMSO = dimethylsulphoxide) and $VOCl_2.2C_5H_5N$. The VO^{2+} entity bonds most effectively to the most electronegative atoms, e.g. F, Cl, O, or N, although bonds to S and P are also known. Fluoro complexes are especially stable.

Structures

The structures of a number of oxovanadium(IV) species have been determined by X-ray means. The characteristic coordination polyhedron is the tetragonal pyramid, in which the axial V═O bond has a length of 1·57–1·68 Å, the equatorial V—O bonds are ~0·4 Å longer, the O═V—O angles are ~106°, and the vanadium atom lies ~0·55 Å above the plane formed by the four singly-bonded oxygen atoms. In some complexes, octahedral coordination is completed by the formation of a further relatively long bond *trans* to the V═O bond (usually ~2·25 Å long). The known X-ray results on oxovanadium complexes are summarized in Table 14; the complexes are discrete with the exception of $VOSO_4$, $VOMoO_4$ and $VO(OCH_3)_3$, which are polymeric.

Certain of the structures are of particular interest. The complex $VO(H_2O)_5SO_4$ occurs in three forms, one monoclinic and two orthorhombic. A recent redetermination[103] of the structure of the monoclinic form indicates that the VO^{2+} group is bound to four water molecules and to one of the oxygen atoms of the sulphate group. The latter is in the *cis*-position to the VO^{2+} group, not the *trans*-position as originally reported. In aqueous solution, however, the cation present is the $VO(H_2O)_5^{2+}$ ion; the four equatorial water molecules of this ion have a residence time of $1{\cdot}35 \times 10^{-3}$ sec, whereas that of the axial water molecule is probably only $\sim 10^{-11}$ sec, as determined by ^{17}O-n.m.r. work[104, 105].

Tartrate ions form dimeric complexes with the oxovanadium(IV) ion of the type $M_4^I[VO(tart)]_2.xH_2O$. However, the structure, as well as the optical and e.s.r. spectra of the dimer formed with *d*-tartrate ions, is different from that formed with *dl*-tartrate. The racemic dimer is the more stable. The dimeric tetranegative ion in the latter case is centro-

99 T. Allersma, R. Hakim, T. N. Kennedy and J. D. Mackenzie, *J. Chem. Phys.* **46** (1967) 154.
100 J. Selbin, *Chem. Rev.* **65** (1965) 153.
101 J. Selbin, *Coord. Chem. Rev.* **1** (1966) 293.
102 R. J. H. Clark, *The Chemistry of Titanium and Vanadium*, Elsevier, Amsterdam (1968).
103 C. J. Ballhausen, B. F. Djurinskij and K. J. Watson, *J. Am. Chem. Soc.* **90** (1968) 3305.
104 J. Reuben and D. Fiat, *Inorg. Chem.* **6** (1967) 579.
105 K. Wüthrich and R. E. Connick, *Inorg. Chem.* **6** (1967) 583.

TABLE 14. X-RAY RESULTS ON OXOVANADIUM SPECIES

Compound	Space group	Z	V = O Å	V – O Å	V – O (*trans* to V = O) Å	Angle O = V – O
$VO(C_5H_7O_2)_2$ [a b]	$P\bar{1}$–C^1_i	2	1·571	1·955–1·983	—	104·8–108·4
$VO(C_{10}H_9O_2)_2$ [b]	$P2_1/c$–C_{2h}^5	4	1·605	1·949–1·984	—	104·0–106·9
$(NH_4)_2[VO(NCS)_4.H_2O].4H_2O$ [c]	$P2_1/c$–C_{2h}^5	4	1·62	2·04 (V–N)	2·22	97
$VO(H_2O)_4SO_4.H_2O$ [d]	$P2_1/c$–C_{2h}^5	4	1·591	1·983–2·048	2·223	93·8–100·7
$VOSO_4$ [e]	$Pnma$–D_{2h}^{16}	4	1·594	2·005–2·056	2·284	96·4–99·8
$VOMoO_4$ [f]	$P4/n$–C_{4h}^3	2	1·677	1·972	2·588	102·4
$(NH_4)_4[VO$–*dd*–$C_4H_2O_6]_2.2H_2O$ [g]	$P4_32_12$–D_4^8	4	1·60	1·79–2·03	h	98·8–122·1*
$Na_4[VO$–*dl*–$C_4H_2O_6]_2.12H_2O$ [i]	$Pbca$–D_{2h}^{15}	4	1·619	1·902–2·004	j	104·6–109·8
$[VO(C_7H_3O_4N)(H_2O)_2].2H_2O$ [k]	$Pbcn$–D_{2h}^{14}	4	1·591	2·017 and 2·027	2·184 (V–N)	96·0 and 107·4
$VO(OCH_3)_3$ [l]	$P2_1/c$–C_{2h}^5	8	1·51 and 1·57	1·74–2·05	2·25 and 2·30	
$VOCl_2.2N(CH_3)_3$ [m]	$Pnma$–D_{2h}^{16}	4	1·59	2·175 (V–N) 2·25 (V–Cl)	—	88·0–119·7*
$VO(C_{12}H_{18}N_2O_2)$ [n]	$P2_1/a$–C_{2h}^5	4	1·598	1·951 2·055 (V–N)	—	—
$VO(C_{32}H_{34}N_4).C_2H_4Cl_2$ [o]	$P2_1/a$–C_{2h}^5	4	1·63	1·97–2·14 (V–N)	—	—
$Na(Et_4N)[VO(C_{14}H_{10}O_3)_2].2C_3H_7OH$ [p]	$P2_1/c$–C_{2h}^5	4	1·584	1·900 and 1·933 (hydroxyl) 1·970 and 1·971 (carboxyl)	—	112·7 and 114·4 (hydroxyl) 103·8 and 104·5 (carboxyl)

* Trigonal bipyramidal structures.

[a] R. P. Dodge, D. H. Templeton and A. Zalkin, *J. Chem. Phys.* **35** (1961) 55 (acetylacetone complex).
[b] P. K. Hon, R. L. Belford and C. E. Pfluger, *J. Chem. Phys.* **43** (1965) 1323, 3111 (benzoylacetone complex).
[c] A. C. Hazell, *J. Chem. Soc.* (1963) 5745.
[d] C. J. Ballhausen, B. F. Djurinskij and K. J. Watson, *J. Am. Chem. Soc.* **90** (1968) 3305.
[e] P. Kierkegaard and J. M. Longo, *Acta Chem. Scand.* **19** (1965) 1906.
[f] H. A. Eick and L. Kihlborg, *Acta Chem. Scand.* **20** (1966) 722.
[g] J. G. Forrest and C. K. Prout, *J. Chem. Soc.* (*A*) (1967) 1312 (the *dd*–tartrate dimer).
[h] The V–V distance within the dimer is 4·35 Å.
[i] R. E. Tapscott, R. L. Belford and I. C. Paul, *Inorg. Chem.* **7** (1968) 356 (the *dl*–tartrate dimer).
[j] The V–V distance within the dimer is 4·082 Å.
[k] B. H. Bersted, R. L. Belford and I. C. Paul, *Inorg. Chem.* **7** (1968) 1557 (pyridine–2,6–dicarboxylate complex).
[l] C. N. Caughlan, H. M. Smith and K. Watenpaugh, *Inorg. Chem.* **5** (1966) 2131.
[m] J. E. Drake, J. Vekris and J. S. Wood, *J. Chem. Soc.* (*A*) (1968) 1000.
[n] L. J. Boucher, D. E. Bruins, T. F. Yen and D. L. Weaver, *Chem. Communs.* (1969) 363 (N,N′-ethylenebis(l-acetonylethylidenimine complex).
[o] R. E. Pettersen and L. E. Alexander, *J. Am. Chem. Soc.* **90** (1968) 3873 (deoxophylloerythroetioporphyrin complex).
[p] N. D. Chasteen, R. L. Belford and I. C. Paul, *Inorg. Chem.* **8** (1969) 408 (diphenylglycollate complex).

symmetric, and consists of two VO^{2+} units bridged by two tetranegative tartrate ions which have lost the protons from the two hydroxyl oxygen atoms as well as from the two carboxyl oxygen atoms. The dimeric anion closely approximates to C_{2h} symmetry, and the co-ordination about the vanadium atoms to square pyramidal[106] (Fig. 3). In the *dl*-dimeric

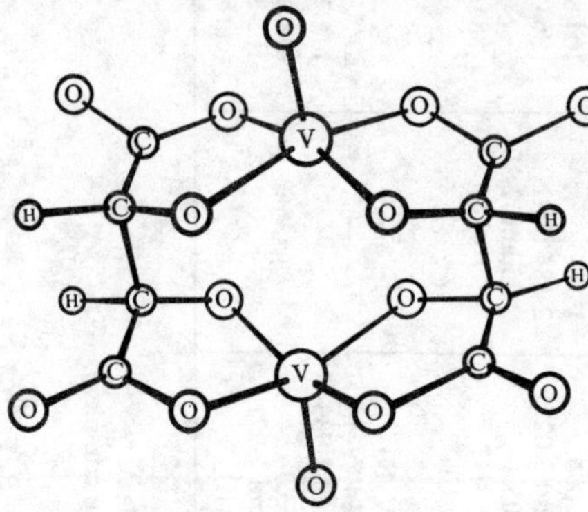

FIG. 3. Structure of the $[VO\text{-}dl\text{-tartrate}]_2^{4-}$ ion[106].

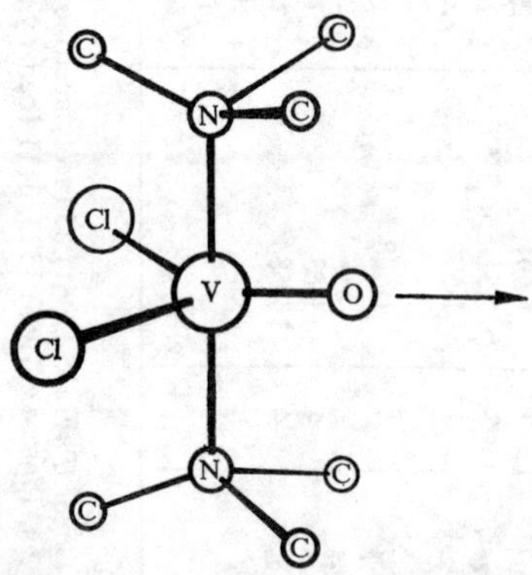

FIG. 4. Structure of the molecule $VOCl_2.2NMe_3$.

anion, the V–V distance is less than in the *dd*-anion, and this difference is associated with a severe distortion from the expected square pyramidal coordination about each vanadium atom; indeed the coordination is best described in this case as distorted trigonal bipyramidal.

The trimethylamine complex $VOCl_2.2NMe_3$ is discrete and surprisingly has a structure (Fig. 4) based on the trigonal bipyramid. It is isomorphous with the vanadium(III) complex $VCl_3.2NMe_3$.

Thermodynamic Properties

The most studied VO^{2+} complex is probably oxo-bis(acetylacetonato-)vanadium(IV), which is monomeric in benzene, and which has a heat of combustion of −1276 kcal/m. Several studies have been made of the reaction

$$VO(acac)_2 + L \rightarrow VO(acac)_2.L$$

[106] R. E. Tapscott, R. L. Belford and I. C. Paul, *Inorg. Chem.* 7 (1968) 356.

in order to determine the heats, entropies and free energies of coordination of ligands (L) to the position *trans* to the V=O bond. The enthalpy decrease on coordination is ~15 kcal/m for oxygen- and nitrogen-donor ligands, but the free energy changes are relatively small[107].

The following thermodynamic data have been obtained for the VO^{2+} ion in aqueous solution[102]: $\Delta H_f^\circ = -113$ kcal/m, $\Delta G_f^\circ = -109$ kcal/m, $\Delta S_f^\circ = -13{\cdot}4$ cal/deg/m.

Electrochemical Properties[102]

At 1 M hydrogen ion concentrations, aqueous vanadium(V) is a moderately good oxidizing agent, the standard reduction potential being $E^\circ = 1{\cdot}000$ V.

$$V(OH)_4^+ + 2H^+ + e \rightarrow VO^{2+} + 3H_2O$$

The potential in alkaline solution is slightly lower.

The VO^{2+}/V^{3+} reduction potential is $E^\circ = 0{\cdot}361$ V at 0°C and 0·337 V at 25°C for the reaction:

$$VO^{2+} + 2H^+ + e \rightarrow V^{3+} + H_2O$$

Solutions of the oxovanadium(IV) ion in the presence of medium or strong acid are stable for months, but they are oxidized in air in the presence of alkalis.

Infrared Spectra

The most characteristic feature of the infrared spectra of oxovanadium(IV) complexes is the very strong, sharp band at 985 ± 50 cm^{-1}. This band is assigned to the V=O stretching frequency[100], and, as expected, lies near the upper frequency limit for those complexes which are known from X-ray work to have the shortest V=O bonds. The V=O stretching frequency for complexes of the type $[VOL_5]^{n\pm}$ falls in the order L = $H_2O > NCS^- > CN^- > DMSO \sim F^-$; this series does not bear any obvious relation to the spectrochemical and nephelauxetic series.

Coordination of a ligand to the sixth octahedral position, i.e. the position *trans* to the V=O bond, brings about a drop of ~50 cm^{-1} in the V=O stretching frequency of the parent complex.

Magnetism and Electron Spin Resonance

The magnetic moments of oxovanadium(IV) complexes lie almost invariably in the range 1·70–1·73 B.M. This result is to be expected because the ligand field about the vanadium(IV) ion (d^1 configuration) is strongly axial, and hence all the orbital contribution to the moment is quenched[102]. The magnetic moments of discrete oxovanadium(IV) complexes should thus be, and indeed are, independent of temperature.

Certain vanadium(IV) complexes, however, show pairwise magnetic exchange similar to that found for d^9 copper(II) complexes. These include a series of 5-substituted N-(2-hydroxyphenyl)salicylideneimine complexes of VO^{2+}, for which the magnetic moments between 1·4 and 300°K have been interpreted[108] in terms of a vanadium–vanadium pairwise interaction with the exchange integral J being in the range 90–218 cm^{-1}.

The e.s.r. g_{av} values for oxovanadium(IV) complexes lie in the range 1·95–2·00, i.e.

107 R. L. Carlin and F. A. Walker, *J. Am. Chem. Soc.* **87** (1965) 2128.
108 A. P. Ginsberg, E. Koubek and H. J. Williams, *Inorg. Chem.* **5** (1966) 1656.

close to the spin-only value of 2·0023. Both this result and the fact that the anisotropy of the g-values is small in comparison with that for the magnetically similar d^9 complexes, indicates that the axial component to the ligand field is high. he Tunpaired electron lies in an orbital of b_2 symmetry.

Optical Spectra

The optical spectra[100] of oxovanadium(IV) complexes are characterized by three *d–d* transitions ($5 < \epsilon < 100$) in the range 8000–32,000 cm^{-1}. The first band ($e \leftarrow b_2$) lies in the range 11,000–14,700 cm^{-1}, the second ($b_1 \leftarrow b_2$) in the range 14,800–20,400 cm^{-1} and the third ($a_1 \leftarrow b_2$) in the range 21,000–31,250 cm^{-1}. The last band is frequently obscured by strong change transfer transitions which typically occur in the blue or ultraviolet spectral regions. The first band in the spectra of some complexes shows vibrational structure (~ 700 cm^{-1} spacings, corresponding to the V=O stretching frequency in the excited state).

Several attempts have been made at detailed interpretations of the ligand field spectra of oxovanadium(IV) complexes. The most successful of these is the molecular orbital treatment of the $VO(H_2O)_5^{2+}$ ion by Ballhausen and Gray[109], in which the bonding scheme is described as follows: a strong σ-bond of a_1 symmetry between the sp_σ oxygen hybrid orbital and the $(4s+3d_{z^2})$ hybrid orbital on the vanadium atom; two π-bonds of e symmetry between the $2p_x$ and $2p_y$ oxygen orbitals and the $3d_{xz}$ and $3d_{yz}$ vanadium orbitals; four σ-bonds between the four equivalent sp_σ (water) oxygen hybrid orbitals and the $(4s-3d_{z^2})(a_1)$, $4p_x$, $4p_y(e)$ and $3d_{x^2-y^2}(b_1)$ vanadium orbitals; the fifth water molecule is bound to the $4p_z(a_1)$ vanadium orbital, leaving the $3d_{xy}(b_2)$ orbital non-bonding and free to hold the single *d*-electron of vanadium(IV). The order of the energy levels is thus $b_2 < e_\pi^* < b_1^* < a_1^*$. The π-bonding in the ion is thus essentially axial, giving rise to a triple bond between the vanadium atom and the unique oxygen atom.

Oxovanadium(V) Species

Vanadium(V) forms two oxo-species, the VO^{3+} and VO_2^+ ions. The former unit is found in the oxyhalides VOX_3 (X = F, Cl, or Br) in which the VO stretching frequencies are at 1058, 1035 and 1025 cm^{-1} respectively. In addition, however, a number of complexes of the type $VOCl_3.L$ and $VOCl_3.2L$ have been characterized in which L can be one of a number of different oxygen- or nitrogen-donor ligands[110, 111]. Such complexes appear to be five- or six-coordinate monomers.

Vanadium oxytrichloride also reacts[110] with ligands containing replaceable hydrogen atoms to produce substitution products of the type $VO(OMe)_3$, $VOCl_2(OMe)$, $VOCl_2(OEt)$, $VOCl(OEt)_2$, $VOCl_2(acac)$, $VOCl(acac)_2$ (Hacac = acetylacetone). Little structural information is available for these complexes. The structure of the alkoxide $VO(OMe)_3$ has, however, been determined (Table 14). The molecule is a linear polymer, with a dimeric repeat unit and alkoxide bridging[102].

The second oxovanadium(V) species, the VO_2^+ ion, is not well characterized, but it may occur discretely in the complexes $VO_2(NO_3)$, VO_2F and $VO_2(SbF_6)$, and as a *cis*-VO_2 unit[112] in the complexes $K_3[VO_2 \text{ oxalate}_2]$ and $K_3[VO_2F_4]$.

109 C. J. Ballhausen and H. B. Gray, *Inorg. Chem.* **1** (1962) 111.
110 H. Funk, W. Weiss and M. Zeising, *Z. anorg. und allgem. Chem.* **296** (1958) 36.
111 H. L. Krauss and G. Gnatz, *Chem. Ber.* **95** (1962) 1023.
112 W. P. Griffith and T. D. Wickins, *J. Chem. Soc.* (*A*) (1968) 400.

2.6. PEROXIDES

A number of different sorts of peroxovanadium complexes have been reported, but few of their structures are known for certain. The best characterized complexes are those of the type K_3MO_8 (M = V, Cr, Nb or Ta), all of which are isomorphous. The chromium salt is known from X-ray structural investigations[113] to contain the chromium atoms each bonded to four peroxo groups in a lateral fashion. The anion has the dodecahedral structure with D_{2d} symmetry and Cr–O bond lengths of 1·846 and 1·944 Å. The transition metal atom is thus eight-coordinate.

In solution, the blue tetraperoxovanadates decompose too fast to permit the measurement of their equivalent conductance. Their infrared spectra are characterized by a strong band at 850 cm^{-1} which has been assigned as the O–O stretching vibration of the co-ordinated peroxo groups[114].

There are also a number of peroxo-complexes in which there are 3, 2 or 1 peroxo-groups per vanadium atom. It has been concluded[115] that in aqueous solution (a) the number of peroxo groups per vanadium atom increases with the alkalinity, (b) increasing acidity increases the degree of polymerization and decreases the number of peroxo groups per vanadium atom and (c) increasing the concentration of hydrogen peroxide decreases the degree of polymerization. The kinetics of formation of the mono- and di-peroxovanadium species in perchloric acid solution have been studied and mechanisms for the reactions proposed[116].

2.7. VANADATES

In aqueous solution, vanadates on acidification undergo a series of complex hydrolysis–polymerization reactions. There are, however, many serious differences in the formulations given to the various polyvanadates in solution, due in part to measurements not being taken under equilibrium conditions (the attainment of which is slow) and in part to invalid reasoning from the experimental data. Several reviews of the subject are available[117, 118], the most recent of which[119] attempts to resolve the various debates.

Solution Studies

Many different techniques have been used to investigate isopolyanions in solution, including those which are concerned with the hydrolysis–polymerization reactions (e.m.f. (pH) measurements, salt cryoscopy, light scattering, equilibrium ultracentrifugation, ion exchange, u.v. and visible spectroscopy) and those which reveal ionic structures or sizes (sedimentation ultracentrifugation, infrared and Raman spectroscopy, n.m.r. spectroscopy, diffusion and dialysis). The various equilibria involved in anion hydrolysis may be formally divided into two categories:

protonation, e.g. $VO_4^{3-} + H^+ \rightleftharpoons HVO_4^{2-}$

condensation, e.g. $2HVO_4^{2-} \rightleftharpoons V_2O_7^{4-} + H_2O$

113 J. D. Swalen and J. A. Ibers, *J. Chem. Phys.* **37** (1962) 17.
114 W. P. Griffith, *J. Chem. Soc.* (1963) 5345; *ibid.* (1964) 5248.
115 J. A. Connor and E. A. V. Ebsworth, *Adv. Inorg. Chem. Radiochem.* **6** (1964) 279.
116 M. Orhanović and R. G. Wilkins, *J. Am. Chem. Soc.* **89** (1967) 278.
117 L. G. Sillén, *Quart. Rev.* **13** (1959) 146.
118 K. F. Jahr, *Angew. Chem.* (*Internat. Ed.*) **5** (1966) 689.
119 M. T. Pope and B. W. Dale, *Quart. Rev.* **22** (1968) 527.

The various vanadate species which have been established are given[119] in Fig. 5 as functions of the total vanadium concentration and the pH. There is little doubt that in strongly alkaline solution, the VO_4^{3-} ion is the predominant species. It is a moderately strong base, hydrolysing in solution to form the HVO_4^{2-} ion. This species will condense to the $V_2O_7^{4-}$ ion and can be protonated to form the $H_2VO_4^-$ ion. Some of the various equilibrium data are summarized in Table 15.

The nature of the species formed by the polymerization of the $H_2VO_4^-$ ion has not been resolved, there being a considerable body of evidence in favour of both the trimer $V_3O_9^{3-}$ as well as of the tetramer $V_4O_{12}^{4-}$.

The nature of the species existing on the acid side of the "metavanadate" region is also

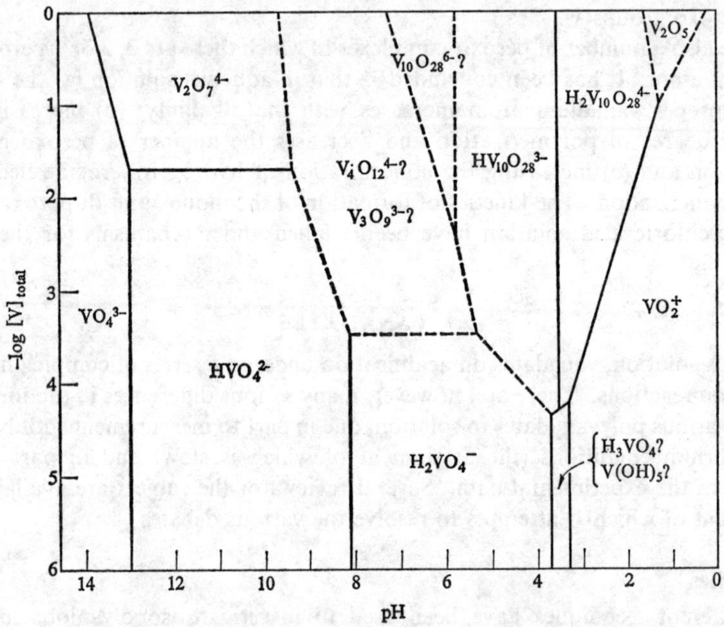

FIG. 5. The approximate conditions of pH and total vanadium concentration under which a given species would be the major solute component of a vanadate solution at 25°C. Demarcation lines about which there is doubt are dashed.[119]

somewhat controversial. Rossotti and Rossotti[120] were able to account for their emf and spectral data on the basis of four species, the VO_2^+, $H_2V_{10}O_{28}^{4-}$, $HV_{10}O_{28}^{5-}$ and $V_{10}O_{28}^{6-}$ ions. However, subsequent work has indicated that the unprotonated decavanadate ion is not a thermodynamically stable species at pH 6·5, as implied by the Rossottis[120]. Brito[121] has concluded from emf measurements at 40°C in the pH range 2–8 that the major vanadate species present are the $V_3O_9^{3-}$, $V_4O_{12}^{4-}$, $HV_6O_{17}^{3-}$, $HV_{10}O_{28}^{5-}$ and $H_2V_{10}O_{28}^{4-}$ ions. He found no evidence for the unprotonated $V_{10}O_{28}^{6-}$ ion.

[120] F. J. C. Rossotti and H. S. Rossotti, *J. Inorg. Nucl. Chem.* **2** (1956) 201; *Acta Chem. Scand.* **10** (1956) 957.

[121] F. Brito, *Acta Chem. Scand.* **21** (1967) 1968.

Structures of Vanadates in Solution

Infrared, Raman, and ^{51}V-n.m.r. work all indicate that the VO_4^{3-} ion is tetrahedral, the four fundamentals occurring[122] at 827(a_1), 340(e), 780(t_2) and 340(t_2) cm^{-1}. Howarth and Richards[123] have shown that when solutions containing the VO_4^{3-} ion are progressively acidified, a plot of the ^{51}V chemical shift against the average charge per vanadium atom is linear for all species up to and including the "metavanadate" ion; there is a discontinuity at the decavanadate ion. This suggests that the HVO_4^{2-}, $HV_2O_7^{3-}$, $V_2O_7^{4-}$ and $V_3O_9^{3-}$

TABLE 15. EQUILIBRIUM DATA FOR AQUEOUS VANADATE SOLUTIONS

Equilibrium	−log K	Method	Temp.	Medium*	Ref.
$HVO_4^{2-} \rightleftharpoons H^+ + VO_4^{3-}$	13·0 ± 0·3	u.v. spect.	25	0·5(NaCl)	a
$VO_4^{3-} + H_2O \rightleftharpoons HVO_4^{2-} + OH^-$	1·01 ± 0·02	u.v. spect.	25	3·0(NaClO$_4$)	b
$2HVO_4^{2-} \rightleftharpoons V_2O_7^{4-} + H_2O$	−1·68 ± 0·04	u.v. spect.	25	3·0(NaClO$_4$)	b
$2HVO_4^{2-} + H_2O \rightleftharpoons HV_2O_7^{3-} + OH^-$	3·18 ± 0·05	e.m.f.	25	0·5(NaCl)	a, c
$HVO_4^{2-} + H_2O \rightleftharpoons H_2VO_4^- + OH^-$	5·89 ± 0·06	e.m.f.	25	0·5(NaCl)	d
$H_2VO_4^- \rightleftharpoons H^+ + HVO_4^{2-}$	7·88	e.m.f. (rapid flow)	20	1·0(NaCl)	e
$HVO_3 \rightleftharpoons VO^{3-} + H^+$	3·70	phase distr.	25	0·5(NaClO$_4$)	f
$H_3VO_4 \rightleftharpoons H_2VO_4^- + H^+$	3·40	e.m.f. (rapid flow)	20	0·1(NaClO$_4$)	e
$VO_2^+ + H_2O \rightleftharpoons HVO_3 + H^+$	3·30	phase distr.	25	0·5(NaClO$_4$)	f
$10VO_2^+ + 8H_2O \rightleftharpoons H_2V_{10}O_{28}^{4-} + 14H^+$	6·75 ± 0·15	e.m.f.	25	1·0(NaClO$_4$)	g
$H_2V_{10}O_{28}^{4-} \rightleftharpoons HV_{10}O_{28}^{5-} + H^+$	4·34	e.m.f. (rapid flow)	20	0·1(NaClO$_4$)	e
$HV_{10}O_{28}^{5-} \rightleftharpoons V_{10}O_{28}^{6-} + H^+$	6·94	e.m.f. (rapid flow)	20	0·1(NaClO$_4$)	e

* This column gives the magnitude of the ionic strength employed and the supporting electrolyte used.

[a] F. Brito and N. Ingri, *Anales Real. Soc. Esp. Fis. Quim.* B**56** (1960) 165.
[b] L. Newman, W. J. LaFleur, F. J. Brousaides and A. M. Ross, *J. Am. Chem. Soc.* **80** (1958) 4491.
[c] N. Ingri and F. Brito, *Acta Chem. Scand.* **13** (1959) 1971.
[d] F. Brito, N. Ingri and L. G. Sillén, *Acta Chem Scand.* **18** (1964) 1557.
[e] A. Schwarzenbach and A. Geier, *Helv. Chim. Acta*, **46** (1963) 906.
[f] D. Dyrssen and T. Sekine, *Acta Chem. Scand.* **15** (1961) 1399.
[g] F. J. C. Rossotti and H. Rossotti, *Acta Chem. Scand.* **10** (1956) 957.

(and/or $V_4O_{12}^{4-}$?) ions are also based on VO_4 tetrahedra, while the decavanadate ion is based on a different coordination.

Both n.m.r. and vibrational spectral work indicate that the $V_3O_9^{3-}$ ion is cyclic in solution[124], although (see below) it has a chain structure in the solid state. The solution data do not distinguish between a cyclic trimer and a cyclic tetramer.

The structure of the $V_{10}O_{28}^{6-}$ ion and its protonated forms, in solution, appear to be uncertain. The two nmr studies that have been made have implied different results, but they suggest that the ion may not have the same structure in solution as it has in the solid state[119] (see below).

122 W. P. Griffith and T. D. Wickins, *J. Chem. Soc.* (*A*) (1966) 1087.
123 O. W. Howarth and R. E. Richards, *J. Chem. Soc.* (1965) 864.
124 W. P. Griffith, *J. Chem. Soc.* (*A*) (1967) 905.

Structures of Vanadates in the Solid State

Structural data on solid vanadates have been reviewed in several places[119, 125, 126]. The V^{5+} ion, on radius-ratio arguments, is considered to be somewhat large for tetrahedral

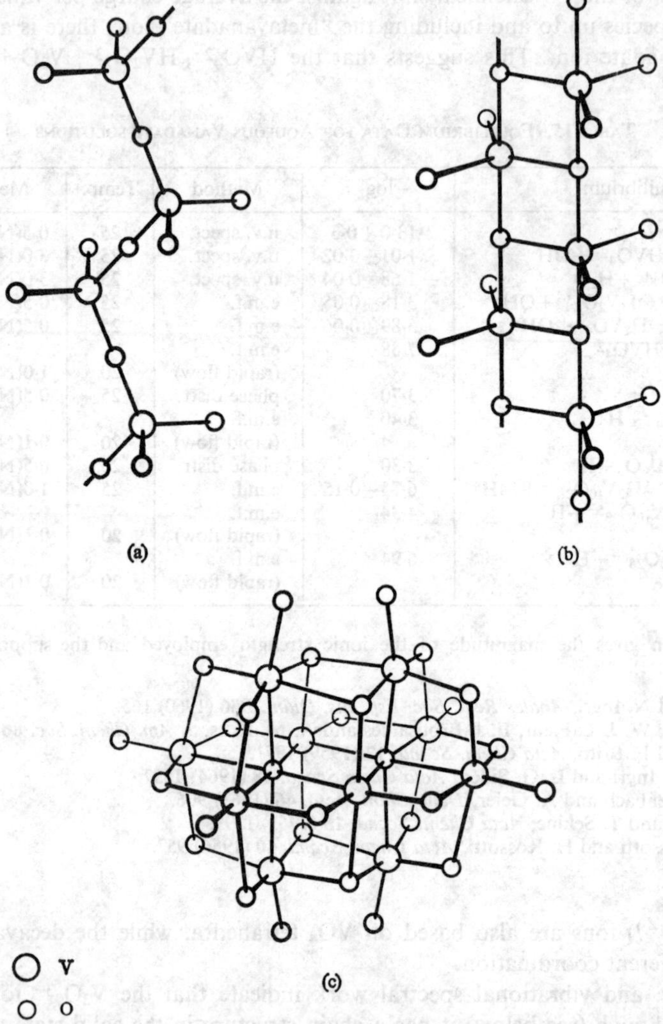

FIG. 6. Structures of (a) infinite chain VO_4 tetrahedra of anhydrous "metavanadates", (b) infinite chain VO_5 polyhedra of hydrated "metavanadates" and (c) the $V_{10}O_{28}^{6-}$ ion.

coordination to oxygen, but somewhat small for octahedral coordination. The possibility of very irregular structures therefore arises. The orthovanadate ion is essentially tetrahedral, although some distortion therefrom may occur; the V–O bond lengths range from 1·64 to 1·74 Å, although a typical value would be 1·66 Å.

125 H. G. Bachmann and W. H. Barnes, *Z. Krist.* **115** (1961) 215.
126 H. T. Evans and S. Block, *Inorg. Chem.* **5** (1966) 1808.

The $V_2O_7^{4-}$ ion is isostructural with the $P_2O_7^{4-}$ ion. From the complete structure determination of the compound $Cd_2V_2O_7$, it emerges that the anion consists of two VO_4 tetrahedra sharing a corner. The bridging oxygen atom has very high thermal motion, and it is thus not clear whether the V–O–V linkage is linear.

No structure is yet known in which the "metavanadate" ion occurs as a discrete unit in the solid state. All structures so far established consist of chains of V–O polyhedra, linked either through corners or edges. Two different types of chain have been described so far[127]. In anhydrous metavanadates (e.g. KVO_3 and NH_4VO_3) the vanadium atoms are tetrahedrally coordinated to four oxygen atoms, two of which are terminal (at 1·66 Å) and two bridging (at 1·81 Å). However, in hydrated metavanadates (e.g. $KVO_3.H_2O$), the vanadium atoms are five-coordinate. There are two terminal oxygen atoms (at 1·65 and 1·67 Å) and three bridging ones (at 1·93, 1·93 and 1·97 Å) arranged about the vanadium atom in an approximate trigonal bipyramid (Fig. 6).

If a vanadate solution is acidified beyond pH 6·5 it turns bright orange and at appropriate concentrations will yield crystalline salts of the decavanadate ions $H_2V_{10}O_{28}^{4-}$, $HV_{10}O_{28}^{5-}$ and $V_{10}O_{28}^{6-}$ in various degrees of hydration, e.g. $10VO_3^- + 4H^+ \rightleftharpoons V_{10}O_{28}^{6-} + 2H_2O$. However, if sufficient acid is added to make the z-value (net anionic charge per metal atom) exactly 0·6 and the solution is then held at 60°C an almost insoluble light-orange compound is formed having the formula $M^IV_3O_8$. If a solution containing slightly less acid is rapidly evaporated at 40°C, dark red crystals of the pentavanadate $M^I_3V_5O_{14}$ are formed. The structures of KV_3O_8 and CsV_3O_8 are of the layer type and consist of highly distorted VO_6 octahedra linked together by sharing corners and edges[126]. Two types of vanadium atom can be distinguished according to the extent of the distortion, which is such that each vanadium atom is displaced in the direction of one oxygen atom. In one case the short V–O distance is 1·624 Å, while in the other it is 1·575 Å; the corresponding long V–O distances are 2·28 and 2·973 Å respectively. Thus the coordination about the vanadium atoms approaches square pyramidal, the equatorial V–O bond lengths being in the range 1·73–1·99 Å. The structure of $K_3V_5O_{14}$ is not known in detail, but the V–O polyhedra are arranged in roughly planar groups of five, three of the polyhedra approximating to square pyramids and two to trigonal bipyramids.[128]

The most interesting vanadate structure is that of the $V_{10}O_{28}^{6-}$ ion, because it provides support for the conclusions of the Rossottis[120] and because the isopolyanion possibly has the same structure in solution as in the solid state. Three independent structural investigations on the compounds $K_2Zn_2V_{10}O_{28}.16H_2O$, $Ca_3V_{10}O_{28}.16H_2O$ and $Na_6V_{10}O_{28}.18H_2O$ have shown[129–131] that in each case the anion is a discrete unit of ten VO_6 octahedra; six of these are condensed into a 2×3 rectangle by sharing of horizontal O–O edges, a further two are inserted symmetrically above and the final two symmetrically below the rectangle, and share sloping O–O edges with the octahedra in the rectangular array. The complex ion thus has D_2 symmetry. The eight outer octahedra are quite strongly distorted towards square pyramidal, the distortion being such that the vanadium atoms are displaced away from the centre of the ion; thus the V–O bond lengths in these octahedra are 1·61 (one), 1·91 (four) and 2·28 Å (one). The two octahedra at the centre of the rectangular sheet are less distorted, the short V–O bond lengths being 1·68 and 1·70 Å (Fig. 6).

127 H. T. Evans, *Z. Krist.* **114** (1960) 257.
128 A. M. Byström and H. T. Evans, *Acta Chem. Scand.* **13** (1959) 377.
129 H. T. Evans, *Inorg. Chem.* **5** (1966) 967.
130 A. G. Swallow, F. R. Ahmed and W. H. Barnes, *Acta Cryst.* **21** (1966) 397.
131 N. Pullman, *Diss. Abs.* **28B** (1967) 40.

The mono- and bi-protonated forms of the decavanadate ion, as well as the recently established compound[132] $(Et_4N)_3H_3V_{10}O_{28}$ probably have structures based upon that described above.

Heteropolyvanadates

The literature on heteropolyvanadates has been summarized by Morette[133]. Recent work has established the existence of the following heteropolyacids: $H_4[PMo_{11}VO_{40}]$, $H_5[PMo_{10}V_2O_{40}]$, $H_6[PMo_9V_3O_{40}]$[134,135] and $H_4[PW_{11}VO_{40}]$, $H_5[PW_{10}V_2O_{40}]$, $H_6[PW_6V_3O_{31}]$[136]. The acid $H_5[PMo_{10}V_2O_{40}]$ is especially stable, and it is thought to be an excellent 1:5 electrolyte.

2.8. SULPHIDES, SELENIDES AND TELLURIDES

A large number of chalcogenide phases have been reported, but some (e.g. VS_5) do not seem to be well established. Structural aspects of this subject have been reviewed by Hulliger[137]. The principal sulphides will be described[138]. However, less information is available on the selenides and tellurides, the most important of which is included in Table 16.

V_3S

Two forms of this sulphide are known; one, the α-form, is stable above 950°C and has a body-centred tetragonal structure, whereas the other, the β-form, is stable below 825°C and has a closely related tetragonal structure. The V–S distances lie in the range 2·31–2·47 Å for both forms.

VS

The brown-black monosulphide may be prepared either by hydrogen reduction of the sesquisulphide at 1000°C or by reaction between the metal and sulphur in a sealed silica tube at 1000°C. It has the nickel arsenide structure over the composition range $VS_{1·00}$–$VS_{1·16}$. It is unstable in air.

V_2S_3

This sulphide may be prepared either by the action of H_2S on V_2O_3, VO_2 or V_2O_5 or by treating V_2O_5 with carbon disulphide at 700°C. It is a black, paramagnetic material with a range of homogeneity which is thought to extend from $VS_{1·17}$ to $VS_{1·53}$. On being heated to 850–950°C *in vacuo*, it dissociates into the monosulphide and sulphur. It is oxidized by air and by nitric acid, but unlike VS_4 is insoluble in dilute sodium hydroxide.

VS_4

The tetrasulphide is a polysulphide which may be prepared by heating the sesquisulphide with an excess of sulphur at 400°C. The mixture is then cooled, and the uncombined sulphur removed with carbon disulphide. It is a black powder which occurs in nature as

132 K. F. Jahr, J. Fuchs and R. Oberhauser, *Chem. Ber.* **101** (1968) 482.
133 A. Morette, in *Nouveau Traité de Chimie Minérale*, ed. P. Pascal, Tome XII, Masson, Paris (1958).
134 P. Courtin and F. Chauveau, *Bull. Soc. Chim. France* (1967) 2461.
135 C. J. Hallada, G. A. Tsigdinos and B. S. Hudson, *J. Phys. Chem.* **72** (1968) 4304.
136 M. P. Courtin, *Bull. soc. chim. France* (1968) 4799.
137 F. Hulliger, *Structure and Bonding*, **4** (1968) 83.
138 R. J. H. Clark, *The Chemistry of Titanium and Vanadium*, Elsevier, Amsterdam (1968).

the ore patronite. It dissolves immediately in potassium hydroxide solution, but resists acids which do not oxidize it. It is unstable above 500°C.

The coordination polyhedron about the vanadium atom is intermediate between a square prism and a square antiprism. As the S–S distance within each S_2^{2-} pair is 2·04 Å,

TABLE 16. BINARY VANADIUM CHALCOGENIDES[a]

Chalcogenide	Structural type	Conductance	Magnetism
V_3S	V_3S	metallic	—
V_5S_4	Ti_5Te_4	metallic	—
VS	NiAs	—	paramag.
V_7S_8	NiAs deriv.	—	antiferromag., $T_N \sim 470$°K?
V_3S_4	Fe_3Se_4	metallic	—
V_2S_3	monoclinic	metallic	2·8 B.M. (Curie-Weiss)
V_5S_8	V_5S_8	metallic	—
VS_4	VS_4	semicond.	diamag.
V_2Se	—	metallic	—
V_5Se_4	Ti_5Te_4	metallic	—
$V_{0\cdot9}Se$	NiAs	metallic	antiferromag. $T_N \sim 300$°K
VSe	~NiAs	semicond.?	antiferromag. $T_N = 407$°K
VSe > 550°K	NiAs	metallic	antiferromag. $T_N = 102$°K
V_7Se_8	NiAs deriv.	metallic	antiferromag. $T_N = 505$°K
V_3Se_4	Fe_3Se_4	metallic	—
V_5Se_8	V_5S_8	metallic	—
VSe_2	CdI_2	metallic	1·7 B.M. (Curie-Weiss)
V_5Te_4	~Ti_5Te_4	metallic	paramag.
VTe_{1+}	NiAs	metallic	1·7 B.M. (Curie-Weiss)
V_3Te_4	Fe_3Se_4	metallic	—
V_2Te_3	monoclinic	metallic	ferromag.?
V_5Te_8	V_5S_8?	metallic	—
$VTe_{1\cdot75}$	$NiAs/CdI_2$	metallic	paramag.
$VTe_{1\cdot85-2}$	monoclinic	metallic	antiferromag. $T_N \sim 450$°K
$V_{1\cdot01}Te_2$	orthorhombic	metallic	—

[a] F. Hulliger, *Structure and Bonding*, **4** (1968) 83.

corresponding to a single bond, only four electrons are used in the cation–anion bonds. Since the compound is diamagnetic, it follows that each cation *d*-electron is involved in a V–V bond, a conclusion substantiated by the observed alternation in the V–V distances along the *c*-axis.

2.9. NITRIDES, AZIDES, PHOSPHIDES, ARSENIDES AND ANTIMONIDES

Nitrides[139]

Vanadium nitride is prepared by the action of nitrogen on a mixture of V_2O_3 and carbon at 1250°C, or better by direct reaction between the metal and nitrogen or ammonia at 900–1300°C. It also results from thermal decomposition of $(NH_4)_3VF_6$ at 600°C[140]. The

[139] H. J. Goldschmidt, *Interstitial Alloys*, Butterworths, London (1967).
[140] H. Funk and H. Böhland, *Z. anorg. und allgem. Chem.* **334** (1964) 155.

nitride is a greyish-violet powder m.p. 2050°C, density 6·04 g/cc, which has the NaCl structure (lattice parameter 4·1398 Å)[141]. The range of homogeneity is wide, extending from $VN_{0\cdot71}$ to $VN_{1\cdot00}$. The standard heat of formation of the compound is −52 kcal/m.

Like the corresponding titanium compound, vanadium nitride liberates ammonia in the presence of alkalis, but it is also attacked by boiling nitric acid. Another phase, V_2N, also has a definite range of homogeneity from $VN_{0\cdot37}$ to $VN_{0\cdot50}$.

The Group V nitrides are much less stable than the Group IV ones, and this has ruled out their possible use as ceramic materials.

Pnictides

The binary phosphides of vanadium are greyish-black materials which are metallic in appearance. They are attacked by hot nitric acid, aqua regia and concentrated sulphuric

TABLE 17. BINARY PNICTIDES OF VANADIUM[a]

Pnictide	Structural type	Conductance
V_3P	Ti_3P	metallic
V_2P	—	metallic?
VP	NiAs	metallic[b]
$V_{0\cdot5\ldots1\cdot0}P$	hexagonal	metallic
VP_2	$NbAs_2$	metallic[b]
$V_{3+}As$	Cr_3Si	metallic
$V_{5+}As_3$	W_5Si_3	metallic
V_3As_2	orthorhombic	—
$V_{\sim2}As$	—	metallic[b]
VAs	MnP	metallic[b]
VAs_2	$NbAs_2$	metallic[b]
V_3Sb	Cr_3Si	supercond.
$V_{\sim3}Sb_2$	Ni_2In	metallic
V_5Sb_4	Ti_5Te_4	metallic
V_6Sb_5	rhombohedral	metallic
VSb	NiAs	metallic
VSb_2	$CuAl_2$	metallic
V_3Bi	Cr_3Si	metallic

[a] F. Hulliger, *Structure and Bonding*, **4** (1968) 83.
[b] Paramagnetic.

acid. Little is known about the other pnictides, but some of their properties are summarized in Table 17.

Ternary Compounds

Vanadium tetrachloride reacts with chlorine azide to produce the explosive compound $VCl_4.N_3$. The latter splits off nitrogen in various solvents to give[142] the volatile nitride chloride $Cl_3V{=}N{-}Cl$. This compound, which reacts with both Lewis acids (e.g. $SbCl_5$) and bases (e.g. pyridine), has a crystal structure consisting of Cl_3VNCl molecules linked together as dimers[143]. Each vanadium atom is linked to one nitrogen atom (apical) and to four chlorine atoms (basal) in the form of a tetragonal pyramid. The V–N–Cl group is

[141] G. Brauer and W. D. Schnell, *J. Less-Common Metals*, **6** (1964) 326.
[142] J. Strähle and K. Dehnicke, *Z. anorg. und allgem. Chem.* **338** (1965) 287.
[143] J. Strähle and H. Bärnighausen, *Z. anorg. und allgem. Chem.* **357** (1968) 325.

almost linear. The relevant bond distances are V–N 1·642, N–Cl 1·588, V–Cl (terminal) 2·204 and 2·298, V–Cl (bridging) 2·383 and 2·463 Å. The compound is thus analogous to oxovanadium(V) species.

A vanadium(V) azide may be prepared by treating $VOCl_3$ at 0°C with gaseous chlorine azide. The resulting compound, which is crystalline, hygroscopic and explosive, has the formula[144] $VOCl_2.N_3$. At room temperature in various solvents it slowly liberates nitrogen and chlorine to form the nitride chloride VNCl. The latter is a brown powder which is undoubtedly polymeric, and which is readily hydrolysed by acids and bases.

2.10. CARBIDES, SILICIDES AND BORIDES

Carbides

Vanadium carbide, VC, may be prepared in an electric furnace either by reduction of vanadium oxides with carbon, or by direct reaction of the elements in a high vacuum above 1300°C. It has the sodium chloride structure, and a wide range of homogeneity. A second

TABLE 18. CARBIDES, SILICIDES AND BORIDES OF VANADIUM[a]

Compound	Lattice parameters Å	Z	M.p. °C	$\Delta H^\circ_{f,\ 298\cdot15}$	ρ (X-ray) g/cc
V_2C	$a = 2\cdot902$, $c = 4\cdot577$	1	2200	—	5·665
VC	$a = 4\cdot18$	4	2830	−24·5	5·649
V_3Si	$a = 4\cdot722$	2	(b)	—	5·67
V_5Si_3	$a = 9\cdot429$, $c = 4\cdot757$	4	2150	—	4·8
VSi_2	$a = 4\cdot571$, $c = 6\cdot372$	3	1750	< -7	4·7
V_3B_2	$a = 5\cdot746$, $c = 3\cdot032$	—	~1900	—	5·44
VB	$a = 3\cdot058$, $b = 8\cdot026$, $c = 2\cdot97$	1	~2300 dec.	—	—
V_3B_4	$a = 3\cdot030$, $b = 13\cdot18$, $c = 2\cdot986$	2	~2400	—	5·10
VB_2	$a = 3\cdot006$, $c = 3\cdot056$	1	—	—	—

[a] H. J. Goldschmidt, *Interstitial Alloys*, Butterworths, London (1967).
[b] Peritectic below ~2060°C.

carbide, V_2C, is also known. Both carbides are dark, very hard, chemically resistant substances, nitric being the only mineral acid to attack them in the cold. These properties give them similar uses to those of titanium carbide.

Certain properties of the carbides, silicides and borides of vanadium are listed in Table 18.

Silicides

The silicides have similar physical properties to the carbides, except for being rather more brittle. Their high thermal conductivities assist their thermal shock resistance, and hence they are attractive as high temperature materials. The silicide V_3Si is of interest as a superconductor, as it has the highest superconductive transition temperature known (17°K).

Borides

The properties of the borides are similar to those of the titanium borides.

[144] K. Dehnicke, *J. Inorg. Nucl. Chem.* **27** (1965) 809.

2.11. ALKOXIDES

Tervalent Alkoxides

The only established tervalent alkoxides of vanadium are $V(OMe)_3$ and $V(OEt)_3$. These compounds may be prepared, under argon, by treating the anhydrous trichloride with lithium methoxide in dry methanol. They are green, non-volatile solids, which are undoubtedly polymeric.

Quadrivalent Alkoxides

Tetraalkoxides of vanadium(IV) may be prepared by alcoholysis of the diethylamino compound as follows[145]:

$$VCl_4 + 4LiNEt_2 \rightarrow V(NEt_2)_4 + 4LiCl$$
$$V(NEt_2)_4 + 4ROH \rightarrow V(OR)_4 + 4Et_2NH$$

The primary alkoxides $V(OR)_4$ (R = Me or Et) are brown solids, the n-pentyl analogue is a green solid, but most compounds of this type are dark brown liquids at room temperature. The tetramethoxide is probably a trimer or a tetramer in the solid state.

The secondary tetraalkoxides are dark green solids or liquids which are probably monomeric in benzene. Analogous tertiary alkoxides are also dark green liquids, with the exception of $V(OBu^t)_4$ which is a blue liquid. They are monomeric in benzene and may be distilled under reduced pressure, albeit with some decomposition.

The e.s.r. spectrum[146] of the compound $V(OBu^t)_4$ as a liquid at 30°C gives $g_{av} = 1{\cdot}964$ and as a polycrystalline solid at −196°C gives $g_{\parallel} = 1{\cdot}940$ and $g_{\perp} = 1{\cdot}984$. The spectral parameter 10 *Dq* has the value 13,500 cm^{-1} for this compound.

Many alkoxyhalides of vanadium(IV) are also known, the properties of which have been summarized by Masthoff and coworkers[147].

Quinquevalent Alkoxides

Many alkoxides of the type $VO(OR)_3$ are known[147]. The most studied of these, the methoxy derivative, is a linear polymer[148] the structure of which has been mentioned in section 2.5. Analogous siloxanes, $VO(OSiR_3)_3$, are also known[149]; those with aliphatic substituents are colourless, thermally very stable, and very sensitive to moisture. The corresponding phenyl derivatives have similar properties except that they are, in addition, hydrolytically stable.

2.12. ORGANOMETALLIC COMPOUNDS

Carbonyls

Vanadium Hexacarbonyl

Vanadium hexacarbonyl was first reported in 1960 by two groups of workers[150, 151]. The preparation was carried out in diglyme according to the scheme[151]:

145 D. C. Bradley and M. L. Mehta, *Can. J. Chem.* **40** (1962) 1710.
146 G. F. Kokoszka, H. C. Allen and G. Gordon, *Inorg. Chem.* **5** (1966) 91.
147 R. Masthoff, H. Köhler, H. Böhland and F. Schmeil, *Z. Chem.* **5** (1965) 122.
148 C. N. Caughlan, H. M. Smith and K. Watenpaugh, *Inorg. Chem.* **5** (1966) 2131.
149 F. Schindler and H. Schmidbaur, *Angew. Chem. Intern. Ed.* **6** (1967) 683.
150 R. L. Pruett and J. E. Wyman, *Chem. and Ind.* (*London*) (1960) 119.
151 R. Ercoli, F. Calderazzo and A. Alberola, *J. Am. Chem. Soc.* **82** (1960) 2966.

$$VCl_3+CO+Na \xrightarrow[250\ at]{100^\circ C} [Na(diglyme)_2]^+[V(CO)_6]^-$$
$$\downarrow H_3PO_4$$
$$V(CO)_6 \leftarrow [HV(CO)_6]$$

The product is in the form of blue-green, pyrophoric crystals which may be sublimed but which thermally decompose above 50°C.

Molecular weight and n.m.r. work originally suggested incorrectly that the molecule was dimeric. In fact, it is isomorphous with the monomeric chromium hexacarbonyl and, both in the solid state as well as in solution, it has a magnetic moment close to that expected for one unpaired electron per vanadium atom (the experimental value is 1·81 B.M. at 20°C in toluene or benzene). The molecule is thus unique among simple metal carbonyls in possessing an odd electron. Electron impact studies show that no dimers are present in the gaseous hexacarbonyl.

The infrared-active CO stretching frequency of $V(CO)_6$ is at 1985 cm^{-1} with a width at half peak height of 22 cm^{-1}. From the e.s.r. spectrum of the pure solid ($g = 2{\cdot}063$) and of the compound in a n-pentane glass ($g_{\parallel} = 1{\cdot}981$, $g_{\perp} = 2{\cdot}103$) at 1·3°K it has been concluded that the compound has a $^2B_{2g}$ ground term at this temperature.

Carbonyl Anion

Various salts of the $V(CO)_6^-$ ion may be precipitated from aqueous solution of the $Na(diglyme)_2^+$ salt by addition of the appropriate halide. The infrared-active CO stretching frequency (t_{1u} species) of the $V(CO)_6^-$ ion occurs at 1859 cm^{-1}, i.e over 100 cm^{-1} below that of the neutral hexacarbonyl. This observation has been interpreted in terms of increased metal→ligand π-bonding in the anionic derivative.

The hexacarbonylvanadate ion may also be prepared by electrochemical reduction of a solution of tris(acetylacetonato)vanadium(III) in the presence of carbon monoxide under pressure.

Vanadium hexacarbonyl, in the presence of Lewis bases, readily disproportionates as follows:

$$3V^0 \rightarrow V^{II}+2V^{-1}$$

Such reactions, which take place directly between the hexacarbonyl and the ligand in benzene, have led to the formation of the following complexes[152]: $VL_6^{2+}[V(CO)_6^-]_2$ (L = aniline, acetonitrile, benzonitrile, or *p*-phenylenediamine); $VL_3^{2+}[V(CO)_6^-]_2$ (L = *o*-phenylenediamine, N,N′-diphenylethylenediamine or *o*-phenanthroline) and $VL_4^{2+}[V(CO)_6^-]_2$ (L = α-picoline, γ-picoline, triphenylphosphine oxide or dimethylsulphoxide). The magnetic moments of these complexes (~ 3·8 B.M.) are independent of temperature between 295° and 77°K and are thus consistent with that expected for bivalent d^3 cations in association with diamagnetic anions. A similar disproportionation takes place in liquid ammonia to produce the compound $V(NH_3)_6[V(CO)_6]_2$, but the liberated carbon monoxide takes part in a secondary reaction as follows[153]:

$$2V(CO)_6+CO+4NH_3 \rightarrow 2NH_4[V(CO)_6]+CO(NH_2)_2$$

---

152 W. Hieber, J. Peterhans and E. Winter, *Chem. Ber.* **94** (1961) 2572.
153 H. Behrens and K. Lutz, *Z. anorg. und allgem. Chem.* **354** (1967) 184.

The liquid ammonia reaction in the presence of KCN or NH_4Cl proceeds differently:

$$3V(CO)_6+6KCN \rightarrow K_4[V(CN)_6]+2K[V(CO)_6]+6CO$$
$$3V(CO)_6+2NH_4Cl+6NH_3 \rightarrow [V(NH_3)_6]Cl_2+2NH_4[V(CO)_6]+6CO$$

A similar disproportionation takes place when heptane solutions of vanadium hexacarbonyl and dimesitylenevanadium are allowed to react; the brown product, which forms immediately, has the formula $[(C_6H_3Me_3)_2V][V(CO)_6]$. Thus vanadium hexacarbonyl is a mild oxidizing agent. It also reacts with aromatic hydrocarbons in the absence of light at 35°C to give red crystalline compounds of the type $[\text{arene } V(CO)_4][V(CO)_6]$. These compounds are insoluble in heptane and in water but are soluble in tetrahydrofuran in which their CO stretching frequencies occur in the range 1975–2070 cm^{-1}. The cations have been

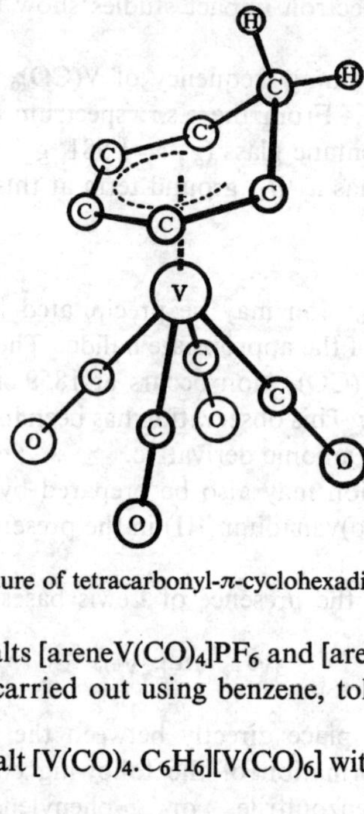

FIG. 7. Structure of tetracarbonyl-π-cyclohexadienylvanadium.

isolated and identified in the salts $[\text{arene}V(CO)_4]PF_6$ and $[\text{arene}V(CO)_4]B(C_6H_5)_4$. Reactions of the above type have been carried out using benzene, toluene, *p*-xylene, mesitylene and many other arenes[154].

Reduction of the benzene salt $[V(CO)_4.C_6H_6][V(CO)_6]$ with sodium borohydride proceeds as follows:

$$[V(CO)_4.C_6H_6][V(CO)_6]+NaBH_4 \rightarrow V(CO)_4.C_6H_7+NaV(CO)_6+\tfrac{1}{2}B_2H_6$$

The complex $V(CO)_4.C_6H_7$ is a red crystalline material which can be sublimed at 40°C/10^{-2} mm and which melts at 66°C with decomposition. On the basis of infrared and n.m.r. evidence it has been shown to be tetracarbonyl π-cyclohexadienylvanadium (Fig. 7). Methyl-substituted cyclohexadienyl compounds $V(CO)_4.C_6H_{7-n}(CH_3)_n$ can also be obtained by the same method. Such compounds are brown-red, moderately stable in dry air in the solid state but rapidly oxidized in solution. They are all monomeric and diamagnetic; their CO stretching frequencies drop with increasing methyl substitution in the ring.

154 F. Calderazzo, *Inorg. Chem.* **3** (1964) 1207; *ibid.* **4** (1965) 223.
155 F. Calderazzo, *Inorg. Chem.* **5** (1966) 429.
156 W. Hieber and E. Winter, *Chem. Ber.* **97** (1964) 1037.

Substitution Products

Triethyl, triphenyl and tri-n-propyl phosphine form red-brown complexes of the type $[V(CO)_4(PR_3)_2]$ on treatment with vanadium hexacarbonyl[156]. These compounds are monomeric and have magnetic moments of 1·79 B.M.; only one CO stretching frequency is infrared active (at $\sim$ 1845 cm^{-1}) and this suggests that the complexes adopt the *trans*-configuration. The complex $[V(CO)_4(PPh_3)_2]$ can be reduced by sodium amalgam in ethanol and the yellow solid $Et_4N[V(CO)_5PPh_3]$ isolated by addition of tetraethylammonium iodide to the reduced solution. Arsine and stibine substitution products are also known, but these are less stable to air than the phosphines.

Tricyclohexylphosphine reacts with vanadium hexacarbonyl to form the yellow dimeric complex $[V(CO)_4(PR_3)_2]_2$. Phosphines of the type RR′PH (R=R′ = C_6H_5; R = C_6H_5, R′ = H; R = R′=H) also form dimeric complexes with vanadium hexacarbonyl, but by a different reaction:

$$2V(CO)_6+2RR'PH \rightarrow [V(CO)_4PRR']_2+H_2+4CO$$

These phosphine complexes are blue-green, non-electrolytic and diamagnetic. Their structures are not known, but they may contain phosphine bridges in which case each vanadium atom is two electrons short of the favoured inert gas structure.

Tetraphenylbiphosphine reacts with vanadium hexacarbonyl to form the blue complex $[V(CO)_4P(C_6H_5)_2]_2$, which is presumably phosphorus bridged. Derivatives of phosphine itself are also known.

Seven-coordinate derivatives of vanadium(0) can be prepared by treating vanadium hexacarbonyl with *o*-phenylenebisdimethylarsine. The resulting light-yellow compound,

TABLE 19. STRUCTURAL DATA ON ORGANO-VANADIUM COMPOUNDS

Compound	Space group	V–C$_{ring}$ Å	V–C$_{carbonyl}$ Å	Colour	μ B.M.
$(\pi\text{-}C_5H_5)V(CO)_4$[a]	*Pnma*	2·28	1·88 and 1·94	orange	diamag.
$(\pi\text{-}C_7H_7)V(CO)_3$[b]	$P2_12_12_1$	2·30	3·10 (V. . .O)	dark brown	diamag.
$(\pi\text{-}C_5H_5)V(\pi\text{-}C_7H_7)$[c]	*Pnma*	2·24	—	purple	1·69
$Ph_3PAuV(CO)_6$[d]		2·68 (V–Au)		yellow-orange	diamag.

[a] J. B. Wilford, A. Whitla and H. M. Powell, *J. Organometal. Chem.* **8** (1967) 495.
[b] G. Allegra and G. Perego, *Ric. Sci. Rend. Sez. A*, **1** (1961) 362.
[c] G. Engebretson and R. E. Rundle, *J. Am. Chem. Soc.* **85** (1963) 481.
[d] H. M. Powell, unpublished work.

$[V(CO)_4diars]_2$ is a dimer and diamagnetic. Treatment of $[Na\ diglyme_2][V(CO)_6]$ with triphenylphosphinegold chloride, Ph_3PAuCl, or with bis(*o*-dimethylarsinophenyl)methylarsinecopper bromide gives[157] the complexes $Ph_3PAu.V(CO)_6$ (yellow-orange) and $[(o\text{-}Me_2AsC_6H_4)_2AsMe]Cu.V(CO)_6$ (light yellow) respectively. These compounds contain V–Au and V–Cu bonds respectively. The crystal structure of the former indicates[158] that the vanadium atom is seven-coordinate, the V–Au bond passing through the centre of one of the VC_3 faces. Important bond lengths for this compound, and for other organo-vanadium compounds for which X-ray structural information is available, are given in Table 19.

[157] A. S. Kasenally, R. S. Nyholm, R. J. O'Brien and M. H. B. Stiddard, *Nature*, **204** (1964) 871.
[158] H. M. Powell, unpublished results.

Nitrosyl Complexes

Vanadium pentacarbonyl nitrosyl is prepared[152] by streaming nitric oxide through a cyclohexane solution of the hexacarbonyl at 0°C. The solution turns violet-red, the compound being identified on the basis of three infrared-active CO stretching frequencies (2108 m, 2064 m and 1992 vs, cm^{-1}) and one NO stretching frequency (1700 vs, cm^{-1}). The nitrosyl compound is very unstable to aerial oxidation.

A second nitrosyl, $[V(CO)_4(NO)PPh_3]$, can be prepared[159] by bubbling nitric oxide through an n-hexane suspension of $[V(CO)_4(PPh_3)_2]$. It is orange, diamagnetic and thought to have the *trans*-configuration on the basis of infrared work; it is more stable in air than the pentacarbonylnitrosyl.

A further nitrosyl, $[(\pi\text{-}C_5H_5)V(CO)(NO)_2]$, has recently been prepared[160] by passing NO through an acetone solution of the complex $Na[(\pi\text{-}C_5H_5)V(CO)_3(CN)]$. The nitrosyl is dark brown and diamagnetic; its CO and NO stretching frequencies in n-hexane are given in Table 20, together with related data on other vanadium carbonyls and carbonyl nitrosyls.

TABLE 20. INFRARED-ACTIVE CO AND NO STRETCHING FREQUENCIES IN CARBONYLS AND CARBONYL NITROSYLS (SOLUTION DATA)

Species	ν(CO)	ν(NO)
$V(CO)_6$[a]	1985	—
$V(CO)_6^-$	1859	—
$V(CO)_5NO$	2108 m	1700 vs
	2064 m	
	1992 vs	
$V(CO)_4(NO)(PPh_3)$	1935 vs	1630 vs
$(\pi\text{-}C_5H_5)V(CO)(NO)_2$	2051 vs	1738 vs
		1650 vs
$V(CO)_4(PPh_3)_2$	1845 vs	—
$V(CO)_4(SbPh_3)_2$	1855 vs	—
$[V(CO)_5(PPh_3)]^-$	1965 m	—
	1861 s	
	1822 vs	

[a] Vapour state.

Cyclopentadienyls

Bis-π-cyclopentadienylvanadium[161]

This compound is formed by treating vanadium tetrachloride with an excess of cyclopentadienylmagnesium bromide or cyclopentadienylsodium with rigorous exclusion of air. If the Grignard reagent is used, the resulting mixture of $(\pi\text{-}C_5H_5)_2V$ and $(C_5H_5)_2Mg$ is sublimed out from the reaction flask, dissolved in ether, and the magnesium compound converted to the insoluble compound biscyclopentadienylcarboxylato magnesium by

159 R. P. M. Werner, *Z. Naturforsch.* **16**b (1961) 478.
160 E. O. Fischer, R. J. J. Schneider and J. Müller, *J. Organometal. Chem.* **14** (1968) 4.
161 E. O. Fischer and S. Vigoureux, *Chem. Ber.* **91** (1958) 2205.

carbon dioxide. The vanadocene can be sublimed out as violet-black crystals in ~50% yield. It can also be prepared by treating vanadium trichloride with cyclopentadienylsodium in dioxan, and by reduction of $(\pi\text{-}C_5H_5)_2VCl_2$ with lithium aluminium hydride.

Vanadocene is comparatively stable thermally, melting at 167–168°C. Its e.s.r. spectrum indicates that the ground state is ^4A ($g = 2{\cdot}00$).

Bis-π-cyclopentadienylvanadium Chloride

This compound can be prepared[162] by direct reaction of VCl_4 with cyclopentadienylsodium. Alternatively a solution of $(\pi\text{-}C_5H_5)_2V$ in ether can be added to an equimolar solution of $(\pi\text{-}C_5H_5)_2VCl_2$ at room temperature. The resulting $(\pi\text{-}C_5H_5)_2VCl$, which is formed rapidly, is purified by sublimation at 165°C/0·2 mm; it forms as indigo-blue crystals, m.p. 206–207°C, in ~70% yield. It reacts with HCl to regenerate the quadrivalent compound $(\pi\text{-}C_5H_5)_2VCl_2$.

The corresponding bromide (m.p. 221–222°C) and iodide (m.p. 214–215°C) are prepared by the following reactions[163, 164]:

$$(\pi\text{-}C_5H_5)_2V + C_2H_5Br \rightarrow (\pi\text{-}C_5H_5)_2VBr + \tfrac{1}{2}C_4H_{10}$$
$$(\pi\text{-}C_5H_5)_2V + \tfrac{1}{2}I_2 \rightarrow (\pi\text{-}C_5H_5)_2VI$$

Bis-π-cyclopentadienylvanadium Dichloride

$(\pi\text{-}C_5H_5)_2VCl_2$ is prepared[165] by treating vanadium tetrachloride with cyclopentadienylsodium or cyclopentadienylmagnesium bromide in tetrahydrofuran. It is a light green compound, m.p. 250°C(dec.), which is insoluble in petroleum ether but soluble in alcohol or water; its magnetic moment is 1·87 B.M. The rather unstable aqueous solution contains the $(\pi\text{-}C_5H_5)_2V^{2+}$ ion, which can be reduced by amalgamated zinc to the purple $(\pi\text{-}C_5H_5)_2V^+$ ion (for which $\mu = 2{\cdot}86$ B.M.).

The compound $(\pi\text{-}C_5H_5)VCl_3$ is also known, and prepared by chlorination of $(\pi\text{-}C_5H_5)_2VCl_2$ in boiling chloroform.

σ-bonded aryls[163, 164]

Treatment of $(\pi\text{-}C_5H_5)_2VCl$ with the stoichiometric quantity of phenyl lithium in dimethoxyethane at −50°C results in the formation of the σ-bonded complex $(\pi\text{-}C_5H_5)_2V(\sigma\text{-}C_6H_5)$ in solution. This compound can be isolated from petroleum ether at −80°C as black crystals, m.p. 92°C.

The same compound can be prepared by treating $(\pi\text{-}C_5H_5)_2VCl_2$ with phenyl lithium. The analogous reaction with $(\pi\text{-}C_5H_5)_2TiCl_2$ proceeds differently, the product being in this case $(\pi\text{-}C_5H_5)_2Ti(\sigma\text{-}C_6H_5)_2$.

Many σ-bonded aryl derivatives of vanadium(III) have been prepared and isolated. Those of the type $(\pi\text{-}C_5H_5)_2VR$ ($R = C_6H_5$, $p\text{-}CH_3C_6H_4$, $p\text{-}(CH_3)_2NC_6H_4$, $C_6H_5CH_2$ or CH_3) melt in the range 90–111°C and are air-sensitive. Their magnetic moments are ~2·7 B.M. over the range 280–90°K.

162 E. O. Fischer, S. Vigoureux and P. Kuzel, *Chem. Ber.* **93** (1960) 701.
163 H. J. de L. Meijer, M. J. Janssen and G. J. M. van der Kerk, *Rec. Trav. Chim.* **80** (1961) 831; *Chem. and Ind.* (*London*) (1960) 119.
164 H. J. de L. Meijer and G. J. M. van der Kerk, *Rec. Trav. Chim.* **84** (1965) 1420.
165 G. Wilkinson and J. M. Birmingham, *J. Am. Chem. Soc.* **76** (1954) 4281.

Organic halides react with $(\pi\text{-}C_5H_5)_2V$ in two distinct ways, where R = benzyl, ethyl or methyl:

$$(\pi\text{-}C_5H_5)_2V + RX\ (excess) \rightarrow (\pi\text{-}C_5H_5)_2VX + \tfrac{1}{2}RR$$
$$2(\pi\text{-}C_5H_5)V\ (excess) + RX \rightarrow (\pi\text{-}C_5H_5)_2VR + (\pi\text{-}C_5H_5)_2VX$$

It seems likely that the initial step in these reactions is the oxidative addition of the organic halide to form $(\pi\text{-}C_5H_5)_2VRX$, although such compounds have not been isolated.

In the purple compound $Li_4[V(C_6H_5)_6]3\cdot5H_2O$, the vanadium atom is apparently σ-bonded to six phenyl groups. It is prepared[166] by treating an ethereal solution of phenyllithium with the complex $VCl_3.3THF$ (THF = tetrahydrofuran), the former being in tenfold excess. The compound has a magnetic moment of 3·85 B.M., loses ether at 50°C and biphenyl at 80°C, and is readily decomposed by acids or water.

Other π-derivatives

Dibenzenevanadium

$(C_6H_6)_2V$ was first prepared by Fischer and Kögler[167] as follows:

$$VCl_4 + Al + 2C_6H_6 \xrightarrow{AlCl_3} [V(C_6H_6)_2]AlCl_4$$

The orange, paramagnetic reaction product is then hydrolysed to give dibenzenevanadium and a derivative of vanadium(V).

The compound forms as brown-red crystals, which, in the absence of air, form stable solutions in most organic solvents and in water. It may be sublimed in high vacuum at 120–150°C, melts at 277–278°C, and decomposes in vacuum at 330°C. Its magnetic moment is 1·73 B.M. and e.s.r. $\langle g \rangle$ value is 1·9866. The latter consists of an eight-line spectrum corresponding to $I = 7/2$ for ^{51}V.

The compound exists in two crystalline forms. Below 18°C, it crystallizes from organic media in a cubic modification, space group $Pa3\text{–}T_h^6$ ($Z = 4$), which is the same as that of dibenzenechromium. Above 22°C, or by sublimation, it forms as monoclinic crystals, space group $P2_1/c\text{–}C_{2h}^5$ ($Z = 2$). The compound is of the sandwich type.

Dimesitylenevanadium has also been prepared; it is a deep red material, m.p. 126–127°C. The general reaction scheme for these preparations is:

$$5V^{I}(arene)_2 \xrightarrow{H_2O} 4V^{\circ}(arene)_2 + V^{v} + 2\ arene.$$

$(\pi\text{-}C_5H_5)V(\pi\text{-}C_7H_7)$

This compound is formed[168] when $(\pi\text{-}C_5H_5)V(CO)_4$ is boiled with cycloheptatriene for 8 hr. It is a purple, sublimable solid, which has a magnetic moment of 1·69 B.M. X-ray work has shown that both rings are symmetrical and parallel. The perpendicular V-ring distances are 1·50 and 1·90 Å for the C_7 and C_5 rings respectively, leading to twelve equal V–C distances of 2·24 Å (Table 19). The compound is more stable in air than dibenzenevanadium.

166 E. Kurras, *Monatsber. Deut. Akad. Wiss. Berlin*, **2** (1960) 109.
167 E. O. Fischer and H. P. Kögler, *Chem. Ber.* **90** (1957) 250.
168 G. Engebretson and R. E. Rundle, *J. Am. Chem. Soc.* **85** (1963) 481.
169 M. C. Baird, G. Hartwell and G. Wilkinson, *J. Chem. Soc.* (*A*) (1967) 2037.

π-CS_2 Complex

Freshly sublimed $(\pi\text{-}C_5H_5)_2V$ reacts immediately with carbon disulphide in a vacuum or in nitrogen to form a deep green, extremely air-sensitive complex[169]. Although this has not been isolated, n.m.r. and infrared evidence indicates that it should be formulated as a π-CS_2 complex of vanadium(IV) (Fig. 8).

Catalytic Properties

In general, organovanadium compounds do not appear to be as effective as organotitanium ones as catalysts for the polymerization of olefins. However, a number are effective, notably the mixture of triphenylaluminium, aluminium tribromide and vanadium tetrachloride[170]. Syndiotactic polypropylene can be prepared using the VCl_4–Et_2AlCl catalyst, and variants on it[171, 172]. Ziegler-Natta catalysis is discussed in the appropriate section under the chemistry of titanium.

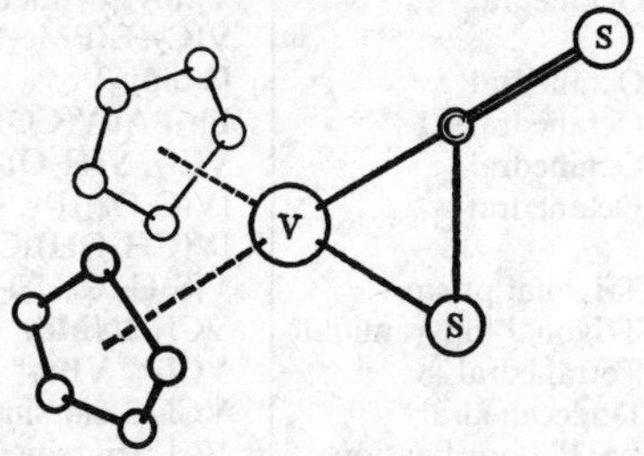

FIG. 8. Structure of bis(π-cyclopentadienyl)(π-carbon disulphide)vanadium(IV).

2.13. COMPLEXES

The complexes of vanadium vary widely in their stability, from those of vanadium(V) which are generally oxidizing to those of vanadium in the bivalent and lower states which are strongly reducing. A summary of the oxidation states and stereochemistries found for vanadium complexes is given[173] in Table 21.

Vanadium –I, 0, I, II

The lowest oxidation states of vanadium are exemplified by the carbonyl $V(CO)_6$ and carbonyl anion $V(CO)_6^-$ (discussed elsewhere), and by tris-bidentate nitrogen- and phosphorus-donor complexes.

2,2′-bipyridyl Complexes

Reduction of an alcoholic or aqueous solution of the complex [V bipy$_3$]I$_2$ with metallic magnesium or zinc produces[174] the zerovalent complex [V bipy$_3$]. A two dimensional X-ray

170 W. L. Carrick, *J. Am. Chem. Soc.* **80** (1958) 6455.
171 G. Natta, I. Pasquon and A. Zambelli, *J. Am. Chem. Soc.* **84** (1962) 1488.
172 J. Boor and E. A. Youngman, *J. Polymer Sci. Part AI*, **4** (1966) 1861.
173 R. J. H. Clark, *The Chemistry of Titanium and Vanadium*, Elsevier, Amsterdam (1968).
174 S. Herzog, *Z. anorg. und allgem. Chem.* **294** (1958) 155.

study of the complex indicates[175] that the vanadium atom is octahedrally coordinated to six nitrogen atoms, with V–N = 2·10 Å and the included N–V–N angle = 73·6°. The compound thus possesses a strong trigonal component to the ligand field.

The compound may be reduced with lithium aluminium hydride in tetrahydrofuran to the complex Li[V bipy$_3$].4THF. Such compounds are all extremely air-sensitive. The colours (in THF solution) and magnetic moments (solid, 293°K) of the whole series of tris-bipyridy

TABLE 21. OXIDATION STATES AND STEREOCHEMISTRIES OF VANADIUM COMPOUNDS

Oxidation state	Coord. no.	Stereochemistry	Examples
−1	6	Octahedral	[Vbipy$_3$]$^{-1}$, [Vphen$_3$]$^{-1}$, $V(CO)_6^-$
0	6	Octahedral	Vbipy$_3$, Vphen$_3$, Vterpy$_2$, $V(CO)_6$, $V[C_2H_4(PMe_2)_2]_3$
1 (d^4)	6	Octahedral	[Vbipy$_3$]$^+$
	7	Octahedral+1	$Ph_3PAuV(CO)_6$
2 (d^3)	6	Octahedral	VCl_2, $V(H_2O)_6^{2+}$, $V(CN)_6^{4-}$
3 (d^2)	6	Octahedral	$[V(C_2O_4)_3]^{3-}$, $VCl_3.3C_4H_8O$, VF_3, $[V(CH_3OH)_4Cl_2]Cl$, $V[OC(NH_2)_2]_6^{3+}$
	6	Trigonal prism	$V[S_2C_2(C_6H_5)_2]_3$[a]
	5	Trigonal bipyramidal	$VCl_3.2NMe_3$
	4	Tetrahedral	VCl_4^-, VBr_4^-
4 (d^1)	8	Dodecahedral	VCl_4.2 diarsine
	7	"4:3" coordination	VCl_4.triarsine
	6	Octahedral	VCl_6^{2-}, VO_2, VCl_4.bipy
	5	Tetragonal pyramid	$VO(acac)_2$, $[VO(NCS)_4]^{2-}$
	5	Trigonal bipyramid	$VOCl_2.2NMe_3$
	4	Tetrahedral	VCl_4, $V(OSiPh_3)_4$
5 (d^0)	8	Dodecahedral	$V(O_2)_4^{3-}$
	6	Octahedral	VF_6^-
	5	Trigonal bipyramidal	VF_5
	4	Tetrahedral	$VOCl_3$, VO_4^{3-}

bipy = 2,2′-bipyridyl; Hacac = acetylacetone; diarsine = *o*-phenylene bisdimethylarsine.
[a] Oxidation state of the metal is not well defined.

complexes with electron configurations t_{2g}^3–t_{2g}^6 are as follows: [V bipy$_3$]$^{2+}$ green, 3·75 B.M., [V bipy$_3$]$^+$ red, 2·80 B.M., [V bipy$_3$] blue, 1·8 B.M., [V bipy$_3$]$^-$ red, 0. Some evidence suggests that, whereas the complex [V bipy$_3$]I exists as such in pyridine solution, in the solid state it is probably a mixed crystal of the [V bipy$_3$] and [V bipy$_3$]$^{2+}$ species.

1,10-phenanthroline Complexes

Many of the synthetic procedures used for the preparation of low-valent bipyridyl complexes of vanadium are not successful with 1,10-phenanthroline. However, the readily prepared dark blue complex [V phen$_3$]I$_2$ may be reduced with dilithium naphthalenide or with dilithium benzophenone in THF as follows[176]:

$$[V\ phen_3]I_2 + Li_2\ benzophenone \rightarrow [V\ phen_3] + benzophenone + 2LiI$$

The neutral complex [V phen$_3$] is thus prepared as dark green crystals which are soluble in

[175] G. Albrecht, *Z. Chem.* 3 (1963) 182.
[176] S. Herzog and U. Grimm, *Z. Chem.* 4 (1964) 32; *ibid.* 6 (1966) 380.

THF, benzene and pyridine to give reddish-brown solutions. Two equivalents of iodine are required to oxidize it back to the original bivalent vanadium complex.

The neutral complex has a magnetic moment at room temperature of 1·90 B.M. consistent with that expected for an octahedrally coordinated ion with the $t_{2g}{}^5$ configuration. Unlike the analogous bipyridyl complex, [V $phen_3$] is not sublimable in high vacuum.

Further reduction of the complex [V $phen_3$] with dilithium benzophenone results in the formation of the corresponding complex in the −1 oxidation state.

$$[\text{V phen}_3]+\text{Li}_2\text{ benzophenone} \xrightarrow{\text{THF}} \text{Li[V phen}_3].3{\cdot}5\text{ THF}+\text{Li benzophenone}$$

This compound crystallizes as black rods which form a dark blue solution in THF. It is highly air sensitive, decomposing under 10^{-4} mm Hg pressure at ~200°C. The very low magnetic moment of the complex (0·63 B.M.) is considered to indicate a spin-paired $t_{2g}{}^6$ configuration (diamagnetic); trace paramagnetic impurities and/or temperature independent paramagnetism account for the low observed moment.

2,2′,2″-terpyridyl Complexes

Terpyridyl, like bipyridyl and phenanthroline, is also capable of forming electron-rich compounds of vanadium[177]. Addition of an alcoholic solution of terpyridyl to an aqueous solution of vanadium(II) sulphate yields a deep green solution, from which green crystals of the complex [V $terpy_2$]I_2 can be precipitated with potassium iodide. The complex, which has a magnetic moment of 3·71 B.M. turns brown in air. If a solution of the complex in dimethylformamide is shaken with magnesium powder, it turns wine-red and black crystals of the neutral complex [V $terpy_2$] can be isolated. This complex has a magnetic moment of 1·73 B.M. and decomposes at 100°C/10^{-3} mm. Other reducing agents such as lithium aluminium hydride are also capable of reducing [V $terpy_2$]I_2 to [V $terpy_2$].

The magnetic data clearly indicate that the bonding scheme for the bipyridyl, phenanthroline and terpyridyl complexes are analogous. However, it is not certain whether these negative oxidation states are mere formalities or whether the complexes are better to be regarded as involving coordinated radical anions.

Phosphine Complex

Tertiary phosphines, because of their capacity to form π-bonds as well as σ-bonds, are well known to stabilize low oxidation states. This is also true with vanadium. Vanadium trichloride in THF in the presence of 1,2-bisdimethylphosphinoethane can be reduced[178] by sodium naphthalenide to the deep brown zerovalent complex $[V(Me_2P(CH_2)_2PMe_2)_3]$. This compound is thermally stable but is oxidized rapidly in air. Its magnetic moment (2·10 B.M.) is indicative of the spin-paired $t_{2g}{}^5$ configuration. The evidence is in favour of the compounds being an octahedrally coordinated monomer.

Cyano-nitrosyl

The orange deliquescent compound $K_5[V(CN)_5(NO)].H_2O$ can be prepared[179] by the

[177] S. Herzog and H. Aul, *Z. Chem.* **6** (1966) 343.
[178] J. Chatt and H. R. Watson, *J. Chem. Soc.* (1962) 2545.
[179] W. P. Griffith, J. Lewis and G. Wilkinson, *J. Chem. Soc.* (1959) 162.

addition of hydroxylamine hydrochloride to potassium vanadate in the presence of an excess of cyanide and hydroxyl ions. The compound is diamagnetic, and its low NO stretching frequency (1575 cm^{-1}) suggests the presence of NO^+ and hence of V^{1-}.

Bidentate Sulphur-Donor Complexes

A number of recent investigations[180, 181] have established the existence of tris-bidentate complexes of the type $[V\,S_6C_6R_6]^n$ where $R = CF_3$ ($n = -2$ or -1), $R = C_6H_5$ ($n = -2$, -1 or 0), or $R = CN$ ($n = -2$ or -1) and $[V(S_2C_6H_3X)_3]^{2-}$ ($X = H$ or CH_3). The precise definition of the oxidation states of the vanadium atoms in these complexes is not certain. It is currently thought that the metal atom is in an oxidation state of two or three and that the ligands are coordinated in the form of radical anions.

X-ray structural work has shown[182] that the complex $V[S_2C_2(C_6H_5)_2]_3$ has a trigonal prismatic structure. The six V–S bond lengths in the complex are 2·338 Å and the internal SVS angles are 81·7°. The five-membered metal chelate rings are approximately planar but the phenyl rings are twisted out of these planes and are apparently not conjugated with them (Fig. 9). The S–S distances (3·064 Å) are relatively short for formally non-bonded atoms and it is thought that S–S interaction may cause some stabilization of the trigonal prism relative to the sterically more favourable octahedron. The compound is purple-black, has a magnetic moment of 1·80 B.M., and $g_{av} = 1{\cdot}990$.

The dinegative anion in the salt $(NMe_4)_2[V(S_2C_2(CN)_2)_3]$ has a magnetic moment of ~1·8 B.M. and a structure intermediate between octahedral and trigonal prismatic stereochemistries. The V–S distances are slightly longer in this salt than in the neutral compound above, averaging[182] 2·36 Å.

Chemistry of Vanadium(II)

Although more extensively investigated than that of titanium(II), the chemistry of vanadium(II) is as yet not well established. In this oxidation state the element closely resembles chromium, but is more readily oxidized.

The best known complex is the violet sulphate, $VSO_4, 6H_2O$, which may be prepared by the electrolytic, sodium amalgam or zinc reduction of a solution of vanadium pentoxide in sulphuric acid. It evolves hydrogen with water but not in acid solution. It is isomorphous with the corresponding zinc(II) salt and thus contains the $V(H_2O)_6^{2+}$ ion[183].

The ammonium, potassium and rubidium double sulphates $M_2V(SO_4)_2.6H_2O$ (Tutton's salts) can likewise be prepared by the electrolytic reduction of a sulphuric acid solution of the pentoxide and the alkali metal sulphate. The amethyst coloured salts have the structure of schönite, $K_2Mg(SO_4)_2.6H_2O$, and hence consist[184] of $V(H_2O)_6^{2+}$ units, the V–O bond lengths being 2·12–2·16 Å. These compounds are rather more stable to oxidation than the simple sulphate, especially in the presence of acid.

180 J. H. Waters, R. Williams, H. B. Gray, G. N. Schrauzer and H. W. Finck, *J. Am. Chem. Soc.* **86** (1964) 4198.
181 A. Davidson, N. Edelstein, R. H. Holm and A. H. Maki, *J. Am. Chem. Soc.* **86** (1964) 2799; *Inorg. Chem.* **3** (1964) 814; *ibid.* **4** (1965) 55.
182 R. Eisenburg and H. B. Gray, *Inorg. Chem.* **6** (1967) 1844; E. I. Stiefel, Z. Dori and H. B. Gray, *J. Am. Chem. Soc.* **89** (1967) 3353.
183 L. F. Larkworthy, J. M. Murphy, K. C. Patel and D. J. Phillips, *J. Chem. Soc.* (*A*) (1968) 2936.
184 H. Montgomery, R. V. Chastain, J. J. Natt, A. M. Witkowska and E. C. Lingafelter, *Acta Cryst.* **22** (1967) 775.

The hexacyanide $K_4[V(CN)_6].3H_2O$ is formed by reduction of vanadium(III) acetate with potassium amalgam in the presence of an excess of potassium cyanide. It is a brown-yellow complex, isomorphous with potassium ferrocyanide; its e.s.r. spectrum has been used to determine the nuclear spin of ^{50}V (six)[185].

The complex chemistry of vanadium(II) has recently been closely investigated[186]. Complexes of the type $VCl_2.4L$ are known with L = H_2O, CH_3CN, C_5H_5N and HCN and of the type $VCl_2.2L$ where L = H_2O, CH_3OH, C_4H_8O, $C_4H_8O_2$, C_5H_5N, $HCON(CH_3)_2$. Most complexes of the first type are blue, and all are considered to be octahedral monomers, their moments being close to, but slightly lower than, the spin only value of 3·89 B.M. for three unpaired electrons; the magnetic moments are independent of temperature, as expected for

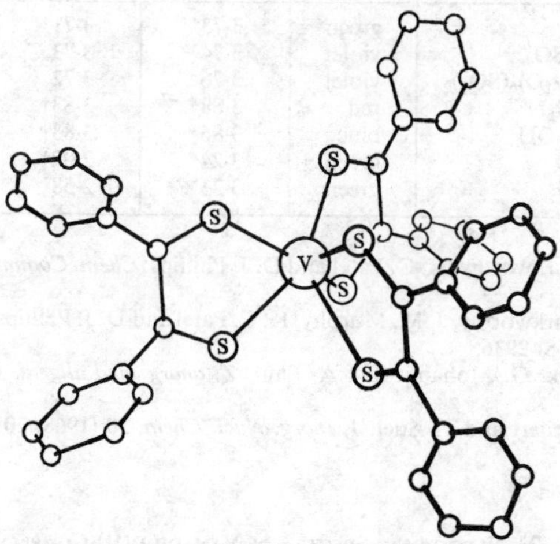

FIG. 9. Structure of the trigonal prismatic complex $V[S_2C_2(C_6H_5)_2]_3$.

the ground $^4A_{2g}$ term (O_h nomenclature). The complexes of the type $VCl_2.2L$ are usually green and their magnetic moments are not only low ($\sim$ 3·2 B.M.) for d^3 complexes, but are also very temperature dependent. These and other properties suggest that such compounds are halogen-bridged polymers. Some properties of typical complexes of vanadium(II) are given in Table 22.

In addition to the above complexes, several others of the type VL_6^{2+} have been isolated in association with the $V(CO)_6^-$ anion. Their properties have been discussed in section 2.12.

The electronic spectra of several octahedrally coordinated vanadium(II) species have been recorded, and the ligand field bands interpreted as follows:

$$\nu_1 = {}^4T_{2g} \leftarrow {}^4A_{2g},\ \nu_2 = {}^4T_{1g}(F) \leftarrow {}^4A_{2g} \text{ and } \nu_3 = {}^4T_{1g}(P) \leftarrow {}^4A_{2g} \text{ (Table 23).}$$

The values of Dq for the various ligands lie in the usual order en > H_2O > CH_3OH > Cl. It is clear that in the LiCl/KCl/VCl_2 melt at 400°C, the principal absorbing species is the VCl_6^{-4} ion. However, as the temperature of the melt is raised to 1000°C the optical

[185] J. M. Baker and B. Bleaney, *Proc. Phys. Soc.* **65**A (1952) 952.
[186] H. J. Seifert and B. Gerstenberg, *Z. anorg. und allgem. Chem.* **315** (1962) 56; H. J. Seifert and T. Auel, *Z. anorg. und allgem. Chem.* **360** (1968) 50; *J. Inorg. Nucl. Chem.* **30** (1968) 2081.

spectrum undergoes a marked change, interpretable in terms of the following octahedral–tetrahedral equilibrium[187].

$$VCl_6^{4-} \rightleftharpoons VCl_4^{2-} + 2Cl^-$$

The high temperature spectrum is similar to that of VCl_2 in Cs_2ZnCl_4 at 900°C, a medium in which tetrahedral VCl_4^{2-} ions would be expected to occur.

TABLE 22. TYPICAL COMPLEXES OF VANADIUM(II)

Complex	Colour	μ(300°K) B.M.	μ(90°K) B.M.	Ref.
$[Ven_3]I_2$	green	3·73	3·71	a
$[V(H_2O)_6]SO_4$	violet	3·74	3·73	b
$(NH_4)_2V(H_2O)_6(SO_4)_2$	violet	3·76	3·72	b
$VCl_2.4C_5H_5N$	red	3·88*	3·83†	c, d
$VCl_2.4CH_3OH$	blue	3·86*	3·83†	d
$VCl_2.2H_2O$	green	3·24*	2·09†	d
$VCl_2.2THF$	green	3·23*	2·58†	d

* 290°K † 82°K

[a] L. F. Larkworthy, K. C. Patel and D. J. Phillips, *Chem. Communs.* (1968) 1667.

[b] L. F. Larkworthy, J. M. Murphy, K. C. Patel and D. J. Phillips, *J. Chem. Soc.* (*A*) (1968) 2936.

[c] H. Funk, G. Mohaupt and A. Paul, *Z. anorg. und allgem. Chem.* **302** (1959) 199.

[d] H. J. Seifert and T. Auel, *J. Inorg. Nucl. Chem.* **30** (1968) 2081.

TABLE 23. ELECTRONIC SPECTRA OF VANADIUM(II) COMPLEXES

Complex	ν_1	ν_2	ν_3	*Dq*	*B*	Method	Ref.
$[V(en)_3]Cl_2.H_2O$	15,600 —	21,500 —	31,700 —	1560	—	C	a
V(II)[MgO]	13,200 —	19,900 —	~30,000 —	1320	—	A	b
$(NH_4)_2[Zn,V(H_2O)_6](SO_4)_2$	12,350 —	18,500 —	27,900 —	1235	—	B	c, d
$(NH_4)_2V(H_2O)_6(SO_4)_2$	12,300 —	18,600 —	28,200 —	1230	—	C	e
$[V(H_2O)_6]SO_4$	11,800 (0·45)	17,500 (0·65)	—	1180	—	B	c, d, f
	11,900 —	17,800 —	~28,000 —	1190	—	C	e
$[V(CH_3OH)_6]Cl_2$	11,700 —	17,450 —	26,520 —	1170	—	B	g
VCl_2	9000 —	14,000 —	21,500 —	900	606	C	h
$(Li,K)_4VCl_6$ (400°C)	7200 (8·1)	12,000 (12·4)	19,050 (17·5)	720	—	D	i

A = crystal transmission spectrum; B = solution spectrum; C = diffuse reflectance spectrum; D = melt spectrum.

[a] L. F. Larkworthy, K. C. Patel and D. J. Phillips, *Chem. Communs.* (1968) 1667.
[b] M. D. Sturge, *Phys. Rev.* **130** (1963) 639.
[c] C. K. Jørgenson, *Acta Chem. Scand.* **12** (1958) 1537.
[d] L. E. Orgel, *J. Chem. Phys.* **23** (1955) 1004.
[e] L. F. Larkworthy, J. M. Murphy, K. C. Patel and D. J. Phillips, *J. Chem. Soc.* (*A*) (1968) 2936.
[f] O. G. Holmes and D. S. McClure, *J. Chem. Phys.* **26** (1957) 1686.
[g] H. J. Seifert and T. Auel, *J. Inorg. Nucl. Chem.* **30** (1968) 2081.
[h] R. J. H. Clark, *J. Chem. Soc.* (1964) 417.
[i] D. M. Gruen and R. L. McBeth *J. Phys. Chem.* **66** (1962) 57.

[187] D. M. Gruen and R. L. McBeth, *J. Phys. Chem.* **66** (1962) 57.

Vanadium(III)

Complexes of vanadium(III) are usually prepared by direct reaction between the vanadium trihalide and the ligand, all operations being carried out on a vacuum line. The complexes are unstable towards air and usually also towards moisture; hence, as for complexes of titanium(III), they are usually sealed into manifolds once prepared. The chemistry of the trichloride has been much more extensively investigated than that of the tribromide and triiodide.

Complexes of the following stoichiometries are known with monodentate ligands (L).

$[VL_6]^{3+}(X^-)_3$, $[VL_4X_2]^+X^-$, $[VL_3X_3]$, $[VL_2X_3]$, $R^+[VX_4L_2]^-$, $(R^+)_2[VX_5L]^{2-}$ and $(R^+)_3[VX_6]^{3-}$
(R = monovalent cation, X = halide ion)

In addition, many complexes are known in which bidentate or multidentate ligands are coordinated to the metal atom. The chemistry of vanadium(III) is closely similar to that of titanium(III).

Complex Ions of the Type ML_6^{3+}

The $V(H_2O)_6^{3+}$ ion is known from crystallographic work to be present as discrete units in the lattice of the caesium alum $CsV(SO_4)_2.12H_2O$. It has also been established as being the principal species present in strong perchloric acid solutions of vanadium(III), on the grounds that the optical spectra of such solutions are identical with the crystal spectrum of the alum[188].

The electronic spectrum of an octahedrally-coordinated vanadium(III) ion should consist of three spin-allowed *d–d* transitions. These can be assigned as follows[173], in terms of O_h nomenclature, the ligand field splitting parameter *Dq*, and the Racah parameter *B*:

$$\nu_1 = {}^3T_{2g} \leftarrow {}^3T_{1g}(F) = 5Dq - 7{\cdot}5B + \tfrac{1}{2}(100Dq^2 + 180Dq.B + 225B^2)^{\frac{1}{2}}$$
$$\nu_2 = {}^3T_{1g}(P) \leftarrow {}^3T_{1g}(F) = (100Dq^2 + 180Dq.B + 225B^2)^{\frac{1}{2}}$$
$$\nu_3 = {}^3A_{2g} \leftarrow {}^3T_{1g}(F) = 15Dq - 7{\cdot}5B + \tfrac{1}{2}(100Dq^2 + 180Dq.B + 225B^2)^{\frac{1}{2}}$$

The energies and extinction coefficients of ν_1 and ν_2 for the $V(H_2O)_6^{3+}$ ion are given in Table 24, together with the analogous data for other vanadium(III) complexes in which there are six equivalent donor atoms[173]. The location of ν_1 and ν_2 allows values for the spectral parameters *Dq* and *B* to be calculated, as indicated above. These data are also included in Table 24. The spectrochemical series derived from the *Dq* values given in Table 24 for the various ligands, namely $CN^- >$ ethylenediamine $> H_2O >$ urea $> F^- > N_3^- > Cl^-$, is similar to that established for Ti(III) and Cr(III).

A number of spectra of vanadium(III) complexes show spin-forbidden bands in addition to the spin-allowed ones. These occur in the 9000–13,000 cm^{-1} region, and have been assigned as ${}^1E_g, {}^1T_{2g} \leftarrow {}^3T_{1g}(F)$. The third spin-allowed band, ν_3, is expected to occur in the 30,000–40,000 cm^{-1} region, and is usually obscured by charge transfer bands.

Complex ions of the type $[VL_6]^{3+}$ are comparatively rare, as is clear from the above discussion. Urea certainly forms an ion of this type[189] but most common ligands do not displace all the halide ions from the first coordination sphere of the parent trihalide. Vanadium trichloride reacts with liquid ammonia at −33°C to give a heterogenous solid with the overall composition $VCl_3.6NH_3$. When this product is washed with liquid ammonia,

[188] H. Hartmann and H. L. Schläfer, *Z. Naturforsch.* **6**a (1951) 754.
[189] G. Barbieri, *Atti Reale Accad. Naz. Lincei, Rend.* **24** (1915) 435.

the composition of the buff residue changes to $VCl_2.NH_2.4NH_3$ ($\mu = 2{\cdot}78$ B.M.) and ammonium chloride is washed out[190]. Vanadium tribromide behaves similarly with liquid ammonia, so that the $[V(NH_3)_6]^{3+}$ ion must be extremely unstable thermally.

Vanadium metal reacts directly with iodine in acetonitrile to form the black complex $[V(CH_3CN)_6](I_3)_3$. This is a rather unstable material, which rapidly loses iodine[191]. It

TABLE 24. ELECTRONIC SPECTRA OF OCTAHEDRAL VANADIUM(III) COMPLEXES WITH SIX EQUIVALENT DONOR ATOMS (cm^{-1})

Complex	ν_1	ν_2	Dq	$15B$	Method	Ref.
$Na_3V(CN)_6$	22,200 (27)	28,600 (50)	2340	7800	A	a
$[V(en)_3]Cl_3$	21,300 (110)	29,500 —	~2150	—	A, B	b
	17,100 (sh)					
$[V(H_2O)_6](ClO_4)_3$	17,250 (5·6)	25,000 (8·3)	1850	9030	A	c, d
$NH_4V(SO_4)_2.12H_2O$	17,800 (3·5)	25,700 (6·5)	—	—	C	e
$CsV(SO_4)_2.12H_2O$	17,800 (4·5)	25,700 (6·5)	—	—	C	e
$(NH_4)_3V(NCS)_6$	16,700 (33)	—	—	—	A	f
$[V(OCN_2H_4)_6](ClO_4)_3$	16,250 (11·9)	24,200 (15·3)	1750	9220	A	d
$K_3[V(C_2O_4)_3]$	16,450 (15·8)	23,500 (sh)	—	—	A	g, h
$K_3[V(CH_2C_2O_4)_3]$	16,450 (21)	23,800 (23)	—	—	A	d, g
$V(III)[Al_2O_3]$77°K	17,510 (∥)	24,930 (∥)	—	—	C	i, j, k
	17,420 (⊥)	25,310 (⊥)	—	—		
K_3VF_6	14,800 —	23,250 —	1610	9730	B	l, m
$(NH_4)_3VF_6$	14,800 —	23,250 —	1610	9730	B	m
$[(n\text{-}C_4H_9)_4N]_3[V(N_3)_6]$	13,300 —	19,400 —	—	—	B	o
$(Li,K)_3VCl_6$(400°C)	11,000 (13·0)	18,000 (35·8)	—	—	D	n
$(pyH)_3VCl_6$(160°C)	11,650 (13·2)	18,350 (30·5)	1265	7700	D	n

Extinction coefficients where available are given in parentheses following the band maxima. A = absorption spectrum in solution; B = diffuse reflectance spectrum; C = crystal transmission spectrum; D = melt transmission spectrum.

[a] J. R. Perumareddi, A. D. Liehr and A. W. Adamson, *J. Am. Chem. Soc.* **85** (1963) 249.
[b] R. J. H. Clark and M. L. Greenfield, *J. Chem. Soc.* (*A*) (1967) 409.
[c] S. C. Furman and G. S. Garner, *J. Am. Chem. Soc.* **72** (1950) 1785.
[d] H. Hartmann and C. Furlani, *Z. phys. Chem.* (Frankfurt), **9** (1956) 162.
[e] H. Hartmann and H. L. Schläfer, *Z. Naturforsch.* **6a** (1951) 754.
[f] S. C. Furman and G. S. Garner, *J. Am. Chem. Soc.* **73** (1951) 4528.
[g] H. Hartmann and H. L. Schläfer, *Angew. Chem.* **66** (1954) 768.
[h] H. Hartmann, C. Furlani and A. Bürger, *Z. phys. Chem.* (Frankfurt), **9** (1956) 62.
[i] W. Low, *Z. phys. Chem.* (Frankfurt), **13** (1957) 107.
[j] M. H. L. Pryce and W. A. Runciman, *Disc. Faraday Soc.* **26** (1958) 34.
[k] D. S. McClure, *J. Chem. Phys.* **36** (1962) 2757.
[l] R. J. H. Clark, *J. Chem. Soc.* (1964) 417.
[m] C. J. Ballhausen and F. Winther, *Acta Chem. Scand.* **13** (1959) 1729.
[n] D. M. Gruen and R. L. McBeth, *J. Phys. Chem.* **66** (1962) 57.
[o] H.-H. Schmidtke and D. Garthoff, to be published.

probably contains the $[VL_6]^{3+}$ cation, as does the complex $[V(CH_3CN)_6]_2(SnCl_6)_3$ but in neither case has this been established beyond doubt.

Some properties of typical vanadium(III) complexes are given in Table 25.

Complex Ions of the Type $VL_4X_2^+$

Although a number of alcohol complexes of the type [V alcohol$_6$]Cl_3 have been reported, these are now known to have been incorrectly formulated. Vanadium trichloride reacts with

[190] D. Nicholls, *Coord. Chem. Rev.* **1** (1966) 379.
[191] B. J. Hathaway and D. G. Holah, *J. Chem. Soc.* (1965) 537.

methanol, ethanol, n- and i-propanol, n-, i- and s-butanol and cyclohexanol to form in solution alcoholates of the type [V alcohol$_4$Cl$_2$]Cl in each case[192]. The complexes isolated

TABLE 25. TYPICAL COMPLEXES OF VANADIUM(III) WITH MONODENTATE LIGANDS

Compound	Colour	μ(B.M.)	T(°K)	Ref.
$(NH_4)V(SO_4)_2 12H_2O$	blue-violet	2·80	300	a
$[V(OC(NH_2)_2)_6]Br_3$	green	2·75	298	b
$[VCl_3.3CH_3CN]$	green	2·79	300	c
$[VCl_3.3C_4H_8O]$	orange	2·80	300	d
$[VCl_3.2NMe_3]$	purple-red	2·72	298	e
$Et_4N[VCl_4.2CH_3CN]$	yellow	2·81	300	c
K_3VF_6	green	2·79	300	f
$(C_5H_6N)_3VCl_6$	purple-pink	2·71	300	g
$(C_5H_6N)_3VBr_6$	orange-brown	2·81	300	g

[a] B. N. Figgis, J. Lewis and F. E. Mabbs, *J. Chem. Soc.* (1960) 2480.
[b] R. L. Carlin and E. G. Terezakis, *J. Chem. Phys.* **47** (1967) 4901.
[c] R. J. H. Clark, R. S. Nyholm and D. E. Scaife, *J. Chem. Soc.* (*A*) (1966) 1296.
[d] R. J. Kern, *J. Inorg. Nucl. Chem.* **24** (1962) 1105.
[e] G. W. A. Fowles and C. M. Pleass, *J. Chem. Soc.* (1957) 1674.
[f] R. S. Nyholm and A. G. Sharpe, *J. Chem. Soc.* (1952) 3579.
[g] G. W. A. Fowles and B. J. Russ, *J. Chem. Soc.* (*A*) (1967) 517.

from the alcohol solutions either have the structural formulae [V alcohol$_4$Cl$_2$]Cl or [V alcohol$_3$Cl$_3$], depending in part on the preparative conditions used.

The electronic spectral bands for some of the alcohol complexes, together with analogous

TABLE 26. ELECTRONIC SPECTRA OF TYPICAL VANADIUM(III) COMPLEXES IN WHICH THERE ARE SIX LIGANDS, NOT ALL EQUIVALENT

Complex	ν_1	ν_2	Dq	$15B$	Method	Ref.
$NH_4[VF_4(H_2O)_2]$	16,000 —	23,800 —	—	—	B	a
$(NH_4)_2[VF_5H_2O]$	15,500 —	23,500 —	—	—	B	a
$[V(CH_3OH)_4Cl_2]Cl$	14,900 (7·9)	22,050 (12·0)	1605	8300	A, B	b, c
$[V(C_2H_5OH)_4Cl_2]Cl$	14,750 (18·5)	21,850 (13·6)	1590	8220	A, B	b, c
$[VF_3.3H_2O]$	16,200 —	24,500 —			B	a
$[VCl_3.3CH_3CN]$	14,400 (17)	21,400 (sh)	1550	8100	A, B	d
$[VCl_3.3C_2H_5CN]$	14,600 (20)	21,750 (sh)	—	—	A, B	d, e
$[VCl_3.3C_4H_8O]$	13,300 —	19,900 —	1400	8300	A, B	c
$[VCl_3.3C_2H_5OH]$	13,900 —	20,850 —	1500	8050	B	c
$Ph_4As[VCl_4.2CH_3CN]$	13,500 (16)	20,000 (37)	1460	8000	A, B	f, g
$Et_4N[VCl_4.2C_5H_5N]$	13,100 —	21,200 —	1430	9300	B	f
$Et_4N[VCl_4.C_{12}H_8N_2]$	{13,800 — {12,500	{22,200 — {20,000	—	—	B	f
$Et_4N[VBr_4.2CH_3CN]$	12,900 —	19,350 —	1395	7450	B	g

Extinction coefficients where available are given in parentheses following the band maxima.
A = solution absorption spectrum; B = diffuse reflectance spectrum.

[a] C. J. Ballhausen and F. Winther, *Acta Chem. Scand.* **13** (1959) 1729.
[b] H. Hartmann and H. L. Schläfer, *Z. Naturforsch.* **6a** (1951) 754.
[c] A. Casey and R. J. H. Clark, *Inorg. Chem.* **8** (1969) 1216.
[d] R. J. H. Clark, J. Lewis, D. J. Machin and R. S. Nyholm, *J. Chem. Soc.* (1963) 379.
[e] M. W. Duckworth, G. W. A. Fowles and R. A. Hoodless, *J. Chem. Soc.* (1963) 5665.
[f] R. J. H. Clark, R. S. Nyholm and D. E. Scaife, *J. Chem. Soc.* (*A*) (1966) 1296.
[g] R. J. H. Clark and A. T. Casey, *Inorg. Chem.* **7** (1968) 1598.

192 A. T. Casey and R. J. H. Clark, *Inorg. Chem.* **8** (1969) 1216.

data for other vanadium(III) complexes involving six non-equivalent ligands, are given in Table 26.

Neutral Complexes of the Type VL_3X_3

Many complexes formed between unidentate ligands and vanadium trihalides are neutral monomers with the formula $[VL_3X_3]$; they are known for L = ethanol, tetrahydrofuran, pyridine and acetonitrile. The properties of some of these complexes are included in Tables 25 and 26. They are invariably very unstable to air and to moisture, and usually suffer solvent exchange when dissolved in solvents other than the parent ligand.

Neutral Complexes of the Type VL_2X_3

Recent work has established that several complexes of vanadium trichloride and tribromide with the formula VL_2X_3 are trigonal bipyramidal. The most studied of these are the isostructural trimethylamine complexes $M(NMe_3)_2X_3$, where M = Ti, V or Cr and X = Cl or Br. An X-ray structural investigation of the complex $TiBr_3.2NMe_3$ has shown that the trimethylamine molecules are in the axial positions[193]. An infrared study of the complexes $VCl_3.2NMe_3$ and $VBr_3.2NMe_3$ has shown that the VN, VCl and VBr stretching frequencies in these complexes occur at 300, 409 and 345 cm^{-1} respectively[194]. The chloro-complex sublimes in vacuum at 100°C.

Triethylphosphine reacts with vanadium trichloride to form[195] the red complex $VCl_3.2PEt_3$, which melts at 160–163°C with decomposition, and which has a magnetic moment of 2·83 B.M. at room temperature. It is a non-electrolyte in acetone, a monomer in benzene, and its electronic spectrum is consistent with a five-coordinate trigonal bipyramidal structure for the molecule. A number of other phosphine and phosphine oxide complexes have the stoichiometry $VCl_3.2L$ and would also appear to be trigonal bipyramidal from their properties. In addition, the sulphides Me_2S, Et_2S and SC_4H_8 react likewise.

The energy level diagram for trigonal bipyramidal d^2 complexes has been worked out, and assignments for the electronic spectra of some of the vanadium(III) complexes of this type have been given[196]. The ground term for such a complex, with the $(e'')^2$ configuration, is ${}^3A_2'$; two spin-allowed transitions (to ${}^3A_1''+{}^3A_2''$ $[e''e']$ and to ${}^3E''$ $[e''e']$) characteristically occur below 8000 cm^{-1}.

Anionic Complexes

Vanadium trichloride reacts with acetonitrile under reflux as outlined above to produce a solution of the green complex $[VCl_3.3CH_3CN]$. This is found[197] to react with large organic halides to produce cream-yellow complexes of the type $R^+[VCl_4.2CH_3CN]^-$ ($R = Et_4N$, $MePh_3As$ or Ph_4As), which behave as 1:1 electrolytes in acetonitrile[197]. The ions $[VCl_3Br.2CH_3CN]^-$ and $[VBr_4.2CH_3CN]^-$ can be prepared similarly[198].

The acetonitrile molecules in the above anions can be replaced[197] by ligands such as

193 M. W. Duckworth, G. W. A. Fowles and P. T. Greene, *J. Chem. Soc. (A)* (1967) 1592.
194 I. R. Beattie and T. Gilson, *J. Chem. Soc.* (1965) 6595.
195 K. Issleib and G. Bohn, *Z. anorg. und allgem. Chem.* **301** (1959) 188.
196 J. S. Wood, *Inorg. Chem.* **7** (1968) 852; P. C. Crouch, G. W. A. Fowles and R. A. Walton, *J. Chem. Soc. (A)* (1968) 2172.
197 R. J. H. Clark, R. S. Nyholm and D. E. Scaife, *J. Chem. Soc. (A)* (1966) 1296.
198 A. T. Casey and R. J. H. Clark, *Inorg. Chem.* **7** (1968) 1598.

pyridine, 2,2′-bipyridyl or 1,10-phenanthroline, (which apparently have stronger donor properties) to produce the ions $[VCl_4.2py]^-$, $[VCl_4.bipy]^-$ and $[VCl_4.phen]^-$. The electronic spectra of these ions (Table 25) are fully consistent with that expected for a d^2 ion coordinated to four chlorine and two nitrogen atoms.

The complex anions VCl_6^{3-} and VBr_6^{3-} were first isolated in 1966 as the pyridinium salts[199]. These complexes are stable in the absence of air, but are solvated by any solvents in which they dissolve. Other properties of these ions are given in Tables 24 and 25.

Melts of KCl/VCl_3 and $NaCl/VCl_3$ are believed to contain the reddish-purple salts K_3VCl_6 and Na_3VCl_6, the melting points of which are 744° and 535°C(dec.) respectively. These melting points are about 20°C lower in each case than those of the corresponding titanium salts[200]. Electronic spectral work by Gruen and McBeth indicates that in a LiCl/KCl eutectic there is a temperature-dependent equilibrium of the following sort:

$$VCl_6^{3-} \rightleftharpoons VCl_4^- + 2Cl^-$$

The tetrahedral form is favoured at higher temperatures[201].

One other anionic chloro species is well established, the $V_2Cl_9^{3-}$ ion, which consists of two octahedra sharing a face. It is isostructural as the caesium salt with the complex $Cs_3Cr_2Cl_9$, the anion of which is known to have D_{3h} symmetry. The electronic spectra and temperature-range magnetism of the $V_2Cl_9^{3-}$ ion has been studied[198, 202]; there is appreciable metal–metal interaction within the dimeric anion.

A number of halo-aquo salts of vanadium are known, including a series of fluoro salts of the type $M_2[VF_5.H_2O]$, M = Tl, NH_4 or Rb. These crystallize in the cubic system and hence it is thought that fluoride and water occupy equivalent, i.e. octahedral, positions about the metal ion[203]. Another series of salts of the type $M(H_2O)_6^{2+}[VF_5.H_2O]$ can be prepared in which M = Co, Ni, Zn or Cd.

Many chloro-aquo salts have been prepared, but their exact formulation has been the subject of much debate. It appears that from hydrochloric acid solutions of vanadium trichloride and alkali metal chlorides (M = Cs, Rb or K) two series of salts can be isolated[204]. At elevated temperatures red compounds of the type $M_2[VCl_5.H_2O]$ are formed but low temperatures favour the formation of green complexes containing more water, e.g. $Cs_2VCl_5.4H_2O$ and $RbVCl_4.6H_2O$.

Several other anionic vanadium species of the type VL_6^{3-} are known, including the $V(CN)_6^{3-}$ and $V(NCS)_6^{3-}$ ions. Although the first was originally reported over 50 years ago, and the spectrum of its supposedly wine-red solution since studied, the solid complex was only isolated in 1966 and then as a *blue* material[190].

The dark red salts $K_3V(NCS)_6.4H_2O$ and $(piperidine)_3V(NCS)_6$ are known, from infrared correlations in the $\nu(CN)$, $\nu(CS)$ and $\delta(NCS)$ vibrational regions, to involve N-bonded thiocyanate groups[205]. Attempts to prepare the analogous pyridinium salt, however, have yielded instead the complex $pyH[V(NCS)_4py_2]$ (cf. the Reinecke salts of chromium(III)). The red selenocyanate salt $[(n\text{-}C_4H_9)_4N]_3[V(NCSe)_6]$ has been prepared,

[199] B. J. Russ and G. W. A. Fowles, *J. Chem. Soc.* (*A*) (1967) 517.
[200] C. Grena, *Bull. Soc. Chim. France*, (1966) 655.
[201] D. M. Gruen and R. L. McBeth, *Plenary Lecture*, *7th Internat. Conf. on Coord. Chem.*, p. 23, Butterworths, London (1963).
[202] R. Saillant and R. A. D. Wentworth, *Inorg. Chem.* **7** (1968) 1606.
[203] L. Passerini and R. Pirani, *Gazz. Chim. Ital.* **62** (1932) 279, 380.
[204] H. J. Seifert and H. W. Loh, *Inorg. Chem.* **5** (1966) 1822.
[205] H. Böhland and P. Malitzke, *Z. anorg. und allgem. Chem.* **350** (1967) 70.

and involves N-bonded NCSe groups. Likewise, a pinkish-brown azide complex $[(n\text{-}C_4H_9)_4N]_3[V(N_3)_6]$ has also been established recently (Table 24).

Complexes with Bidentate Ligands

Vanadium(III) forms complexes with neutral bidentate ligands (e.g. ethylenediamine), ligands with one replaceable proton (e.g. acetylacetone) and ligands carrying two negative charges (e.g. oxalate). Many of the complexes have exact analogues in titanium(III) and chromium(III) chemistry.

With neutral bidentate ligands(B) complexes of two types are formed. The first of these, exemplified by ethylenediamine and propylenediamine[206], has the formula $[VB_3]Cl_3$; these are red-brown complexes with magnetic moments of 2·80 B.M. at room temperature. On the other hand, 2,2′-bipyridyl reacts with vanadium trichloride to form yellow-brown complexes of the type $[VB_2Cl_2]Cl$ and $[VB_2Cl_2][VBCl_4]$.

Neutral tris-bidentate complexes of vanadium(III) are known for many ligands containing a single replaceable proton, e.g. acetylacetone, benzoylacetone, dibenzoylmethane and quinoline. The structures of several tris-β-diketone complexes have been established by n.m.r. techniques[207], both *cis*- and *trans*-isomers of a given tris-chelate complex being detected

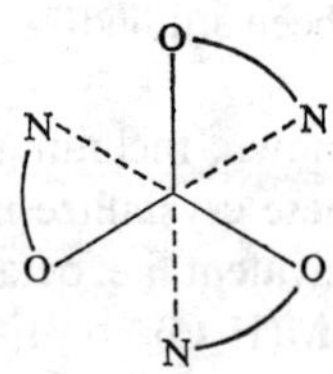

cis

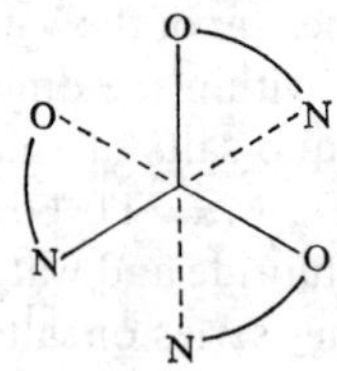

trans

in solution in those cases for which the ligands are not very unsymmetrical sterically. For further series of tris-β-ketoamine, tris-salicylaldehyde and tris-salicylaldimine vanadium(III) complexes similar n.m.r. work demonstrates unequivocally the *trans* stereochemistry for all species[208].

Oxalate, malonate and other bidentate ligands carrying two negative charges form complexes of the type $K_3[V\ ox_3]$ and $K_3[V\ mal_3]$ with vanadium(III). Some spectroscopic data on these complexes are included in Table 24; their magnetic moments of 2·80 B.M., independent of temperature, indicate the presence of a strong trigonal field in the complex anions, already implicit in their D_3 symmetry.

Acetate and benzoate ions form complexes with vanadium(III) of a further type. On treating vanadium diboride with anhydrous acetic or benzoic acids at 110°C for 2 weeks in a sealed tube, the following reaction takes place:

$$2VB_2+6RCO_2H \rightarrow V_2(O_2CR)_6+3H_2+4B$$

The product is a light green powder, with a low magnetic moment (0·77 B.M.); its structure is considered to consist of two vanadium atoms, each bonded to one terminal carboxylate

206 R. J. H. Clark and M. L. Greenfield, *J. Chem. Soc.* (*A*) (1967) 409.
207 F. Röhrscheid, R. E. Ernst and R. H. Holm, *Inorg. Chem.* 6 (1967) 1315.
208 F. Röhrscheid, R. E. Ernst and R. H. Holm, *Inorg. Chem.* 6 (1967) 1607; *J. Am. Chem. Soc.* **89** (1967) 6472.

group and to four bridging carboxylate groups[209], i.e. the structure is analogous to that of the well-known copper acetate hydrate dimer.

Complexes with Multidentate Ligands

Two triarsines, methylbis(*o*-dimethylarsino-phenyl)arsine (abbreviated *o*-TAS) and tris-1,1,1-dimethylarsinomethyl)ethane (*v*-TAS), are known to react with vanadium trichloride in tetrahydrofuran under nitrogen to form reddish-brown complexes of the type $[VCl_3.TAS]$. Infrared and other evidence indicates that all three arsenic and all three chlorine atoms are bonded to the metal atom, and hence that the complexes are six-coordinate neutral monomers. The VCl stretching frequencies ($\sim 330\ cm^{-1}$) indicate that the chlorine atoms in both complexes occupy mutually *cis*-positions[210]. Chromium trichloride forms analogous complexes.

The terdentate nitrogen-donor ligand, 2,2′,2″-terpyridyl forms a yellow-brown 1:1 adduct with vanadium trichloride which is probably a six-coordinate monomer. EDTA complexes of vanadium(III) are also known, but their structures are uncertain.

Tetrahedral Complex Ions

On heating the six-coordinate vanadium(III) complex $Ph_4As[VCl_4.2CH_3CN]$ to 80–100°C in a vacuum, it readily loses the two molecules of acetonitrile, and undergoes a profound colour change from yellow to deep-blue. That the resulting complex, Ph_4AsVCl_4, contains a tetrahedral VCl_4^- ion is established beyond doubt by the isomorphism between this complex and the corresponding salt $Ph_4AsFeCl_4$, which is known from X-ray work to contain the tetrahedral $FeCl_4^-$ ion. Similarly, thermal decomposition of the complex $Et_4N[VBr_4.2CH_3CN]$ produces Et_4NVBr_4, which also contains a tetrahedral anion. These salts are highly unstable to air, and readily solvate in any solvent in which they dissolve[198].

The electronic spectra of these anions differ, as expected, from those of octahedral vanadium(III) complexes. For the VCl_4^- ion, two band systems have been located at ~ 9000 and $\sim 15{,}000\ cm^{-1}$. These are assigned as the $^3T_1(F) \leftarrow {}^3A_2$ and $^3T_1(P) \leftarrow {}^3A_2$ spin-allowed transitions. An analysis of the spectra of the VCl_4^- and VBr_4^- ions indicates that the values of the ligand field splitting parameter $10Dq$ for these ions are 5530 and 5200 cm^{-1} respectively. The Dq value for four tetrahedrally-coordinated halogens is thus less than half that of six octahedrally coordinated halogens (cf. Table 23).

Tetrahedral vanadium(III) electronic spectra have also been recorded for vanadium(III) inserted in a $CsAlCl_4$ lattice, in a $KAlCl_4$ melt at 700°C, and in the CdS (wurtzite) structure. The extinction coefficients of the absorption bands are usually ~ 200, much higher than is typical for the spectra of octahedral vanadium(III) species[173].

Magnetic Properties of Vanadium(III) Complexes

The magnetic moments of octahedral vanadium(III) complexes ($^3T_{1g}$ ground term) are expected to be $\sim 2{\cdot}8$ B.M. at room temperature and to fall appreciably with a lowering of the temperature[211]. In practice the temperature dependence of these magnetic moments is less marked than is expected for an orbital triplet ground term; indeed, it is consistent with a splitting of the ground triplet by some 500–2000 cm^{-1}. The splitting is most obviously brought about by a non-equivalence of all six ligands.

[209] N. N. Greenwood, R. V. Parish and P. Thornton, *J. Chem. Soc.* (*A*) (1966) 320.
[210] R. J. H. Clark, M. L. Greefield and R. S. Nyholm, *J. Chem. Soc.* (*A*) (1966) 1054.
[211] B. N. Figgis and J. Lewis, *Progr. Inorg. Chem.* **6** (1964) 37.

Tetrahedral vanadium(III) complexes, with a $^3A_2$ ground term, are expected to give rise to a temperature independent magnetic moment of $\sim 2{\cdot}7$ B.M. and do so in practice.

Vanadium(IV)

Complexes of Vanadium Tetrafluoride

Little research has yet been done on the chemistry of vanadium tetrafluoride. Indeed such adducts of VF_4 which are known are often best made by reaction between the pentafluoride and an excess of the ligand. For instance[212], the pentafluoride with ammonia reacts to form an unidentified orange-brown solid which may be a simple adduct; however this

TABLE 27. TYPICAL COMPLEXES OF VANADIUM(IV)

Compound	Colour	μ(B.M.)20°C	M.p. °C	Ref.
$VF_4.NH_3$	buff	1·83	250 dec.	a
$VF_4.py$	grey-pink	1·79	150 dec.	a
$VCl_4.bipy$	brown	1·77	—	b
$VCl_2(acac)_2$	dark blue	—	160–165	c
$VCl_4(EtS(CH_2)_2SEt)$	black	1·69	130 dec.	d
K_2VF_6	pink-yellow	1·73	—	e
Cs_2VCl_6	dark red	1·75	—	f

[a] R. G. Cavell and H. C. Clark, *J. Chem. Soc.* (1962) 2692.
[b] R. J. H. Clark, *J. Chem. Soc.* (1963) 1377.
[c] H. Funk, G. Mohaupt and A. Paul, *Z. anorg. und allgem. Chem.* **302** (1959) 199.
[d] R. J. H. Clark and W. Errington, *Inorg. Chem.* **5** (1966) 650.
[e] W. Liebe, E. Weise and W. Klemm, *Z. anorg. und allgem. Chem.* **311** (1961) 281.
[f] P. A. Kilty and D. Nicholls, *J. Chem. Soc.* (1965) 4915.

material loses ammonia and then ammonium fluoride to yield a buff complex $VF_4.NH_3$. Pyridine likewise reacts with the pentafluoride to yield a vanadium(IV) complex, VF_4py. Both of these complexes are insoluble in common organic solvents, and are doubtless fluorine-bridged polymers. Their magnetic moments ($\sim 1{\cdot}8$ B.M.) indicate the presence of vanadium(IV). The corresponding reactions of niobium pentafluoride with these ligands(L) yield stable adducts of the type $NbF_5.2L$; this result reflects the greater stability of the highest oxidation state of a second-row transition metal in this part of the periodic table relative to that of a first-row transition metal. The vanadium(IV) complexes can also be prepared by direct reaction between the tetrafluoride and the ligand[212].

Ethylenediamine also reduces the pentafluoride the product being the complex Ven_3F_4. The insolubility of this material in common solvents has precluded the determination of its structure. Some properties of vanadium(IV) complexes are given in Table 27.

Complexes of Vanadium Tetrachloride

The reactions of vanadium tetrachloride with σ-donor ligands are closely similar to those of titanium tetrachloride, but they have been less extensively investigated. The complexes appear, in general, to be six-coordinate monomers although in few cases has this been established because of their low solubilities in suitable solvents and their great ease of hydrolysis. The usual hydrolysis products of these complexes are the very stable oxo-

[212] R. G. Cavell and H. C. Clark, *J. Chem. Soc.* (1962) 2692.

vanadium(IV) species, the chemistry of which has already been discussed. No single crystal X-ray study of any adduct of the tetrachloride has yet been carried out, although the structure of the complex VCl_4.2 diarsine is known[213] owing to its isomorphism with the known eight-coordinate titanium complex $TiCl_4$.2 diarsine.

Vanadium tetrachloride forms 1:2 adducts VCl_4.2L with the ligands acetonitrile, pyridine, phosphorus oxytrichloride, aldehydes, pyrazine, 2,6-dimethyl pyrazine, benzophenone, tetrahydrofuran, tetrahydropyran, substituted pyridine N-oxides and many aromatic amines[214, 216]. In general these compounds are reddish-brown and have magnetic moments close to 1·7 B.M. With some ligands, such as pyridine, dialkylsulphides, alkylcyanides or tetrahydrofuran, if present in excess, the vanadium is reduced to the tervalent state and complexes of the type VCl_3.3L or VCl_3.2L are formed. However, in other cases, for example with triphenylphosphine, the reduction is not complete and a mixture of complexes in different oxidation states is formed.

Although tertiary amines react in excess with vanadium tetrachloride to form the tervalent derivatives $VCl_3.2NR_3$, reaction with primary or secondary amines in excess yields poorly-defined aminolysis products of quadrivalent vanadium of the types $VCl_3(NHR)_2.xNH_2R$ and $VCl_2(NR_2)_2.yNHR_2$.

The tetrachloride also reacts with NO to form the dark purple paramagnetic compounds[217] VCl_4.NO, V_2Cl_7.NO and $V_2Cl_8(NO)_5$ and the brown diamagnetic compound[218] $VCl_2(NO)_3$. They all liberate NO in the presence of moisture, but none has been adequately characterized.

With typical bidentate ligands(B), e.g. 2,2′-bipyridyl, 1,10-phenanthroline, 1,2-dimethoxyethane and *o*-phenylenebisdiethylarsine, vanadium tetrachloride forms[219] brown, paramagnetic complexes of the type VCl_4.B. Some of these, notably the bipyridyl and phenanthroline adducts, hydrolyse to well-defined products of the type $VOCl_2$.B, but in other cases the metal–ligand bond is also broken during the hydrolysis.

Vanadium tetrachloride thus forms adducts with nitrogen-, phosphorus-, arsenic-, oxygen- and sulphur-donor ligands. It is also known to form a purple-grey adducts with the bidentate selenium-donor ligands $CH_3Se(CH_2)_nSeCH_3$ ($n = 2$ or 3).

In addition to all the above addition reactions, many substitution reactions also take place between vanadium tetrachloride and suitable ligands, i.e. one with a replaceable proton. For instance[220] the tetrachloride reacts rapidly and exothermically with alcohols in cold benzene to give dichloride dialkoxide alcoholates $VCl_2(OR)_2$.ROH (R = Me, Et, n-Pr, i-Pr, n-Bu, i-Bu and s-Bu). The structures, and indeed in some cases the formulae, of these complexes remain in doubt[214]. With acetylacetone and benzoylacetone vanadium trichloride forms the compounds $[VCl_2(C_5H_7O_2)_2]$ and $[VCl_2(C_{10}H_9O_2)_2]$ in direct analogy with the behaviour of titanium tetrachloride[214].

Complete replacement of the halogen atoms of the tetrachloride can be brought about by treatment with lithium dialkylamides to give[221], for example, the dark green liquid $V(NEt_2)_4$.

213 R. J. H. Clark, J. Lewis and R. S. Nyholm, *J. Chem. Soc.* (1962) 2460.
214 H. Funk, G. Mohaupt and A. Paul, *Z. anorg. und allgem. Chem.* **302** (1959) 199.
215 P. A. Kilty and D. Nicholls, *J. Chem. Soc.* (1965) 4915.
216 M. W. Duckworth, G. W. A. Fowles and R. A. Hoodless, *J. Chem. Soc.* (1963) 5665.
217 A. G. Whittaker and D. M. Yost, *J. Am. Chem. Soc.* **71** (1949) 3135.
218 W. Beck, K. Lottes and K. Schmidtner, *Angew. Chem.* (*Intern. Ed.*) **4** (1965) 151.
219 R. J. H. Clark, *J. Chem. Soc.* (1963) 1377; *ibid.* (1965) 5699.
220 D. C. Bradley, R. K. Multani and W. Wardlaw, *J. Chem. Soc.* (1958) 4647.
221 D. C. Bradley and M. L. Mehta, *Can. J. Chem.* **40** (1962) 1183; I. M. Thomas, *Can. J. Chem.* **39** (1961) 1386.

The latter reacts vigorously with alcohols to form the tetraalkoxide $V(OEt)_4$. E.s.r. studies of both of these compounds have been made[222].

The electronic spectra of some typical vanadium(IV) complexes are given in Table 28. The $^2E_g \leftarrow {}^2T_{2g}$ transition is usually asymmetric, as in the case of titanium(III) spectra, but lies to higher energy than in the latter.

Hexahaloanions

The compounds M_2VF_6 (M = K, Rb or Cs) are best prepared by heating the tervalent salts K_2VF_5 with 1% fluorine gas (diluted with CO_2) at ~ 500°C for 10–15 min. The products are then cooled rapidly to −78°C to avoid the formation of any significant amount of vanadium(V). These compounds are pinkish-yellow, stable in dry air, and have magnetic moments of ~ 1·7 B.M. at 295°K[223]. Their lattice parameters are available.

TABLE 28. ELECTRONIC SPECTRA OF SOME OCTAHEDRAL VANADIUM(IV) COMPLEXES

Complex	$^2E_g \leftarrow {}^2T_{2g}$ (cm^{-1})		Splitting	Method	Ref.
$VCl_4.C_{10}H_8N_2$	21,300	17,400 (sh)	3900	A	a
$VCl_4.C_{12}H_8N_2$	21,700	18,000 (sh)	3700	A	a
$VCl_4.(CH_3OCH_2-)_2$	20,400	18,500	1900	A	b
$VCl_4.(CH_3SCH_2-)_2$	18,300	17,400	900	A	b
$VCl_4.(C_2H_5SCH_2-)_2$	18,500	17,500	1000	A	b
$VCl_4.2C_5H_5N$	~15,750	—	—	A	c
K_2VF_6	20,100	—	—	C	d
$(pyH)_2VCl_6$	15,400 (89)	asym.	—	A, B	c

Extinction coefficients in parentheses.
$C_{10}H_8N_2$ = 2,2′-bipyridyl; $C_{12}H_8N_2$ = 1,10-phenanthroline;
A = diffuse reflectance spectrum; B = solution spectrum in CH_3CN;
C = KCl pellet spectrum.

[a] R. J. H. Clark, *J. Chem. Soc.* (1963) 1377.
[b] R. J. H. Clark and W. Errington, *Inorg. Chem.* **5** (1966) 650.
[c] P. A. Kilty and D. Nicholls, *J. Chem. Soc.* (1965) 4915.
[d] H. D. Bedon, W. E. Hatfield, S. M. Horner and S. Y. Tyree, *Inorg. Chem.* **4** (1965) 743.

The hexachloro salts M_3VCl_6 (M = pyridinium, quinolinium or isoquinolinium) can be prepared by an interesting reaction involving the fission of the V–O bond of the green oxotetrachlorovanadates by thionyl chloride[215].

$$M_2VOCl_4 + SOCl_2 \rightarrow M_2VCl_6 + SO_2$$

The caesium, potassium and ammonium salts of the VCl_6^{2-} ion can be prepared by a similar procedure using nitromethane as solvent, and the diethyl- and triethylammonium salts by direct reaction of the alkylammonium halide with vanadium tetrachloride in chloroform.

The hexachlorovanadates(IV) are extremely sensitive to moisture but are otherwise stable up to at least 50°C. At 130°C the pyridinium salt evolves hydrogen chloride to form principally the adduct VCl_4.2pyridine. All the salts have magnetic moments near 1·74 B.M. at 293°K. The value of the ligand field splitting parameter (10*Dq*) for the VCl_6^{2-} ion is

222 C. E. Holloway, F. E. Mabbs and W. R. Smail, *J. Chem. Soc.* (*A*) (1968) 2980.
223 W. Liebe, E. Weise and W. Klemm, *Z. anorg. und allgem. Chem.* **311** (1961) 281.

15,400 cm^{-1}, some 35% greater than for the isoelectronic $TiCl_6^{3-}$ ion; this increase in *Dq* for a given set of ligands with an increase in the oxidation state of the metal ion is general.[224]

Salts of the ions VBr_6^{2-} and VI_6^{2-} have not been reported.

Vanadium(V)

As indicated in the previous section, common ligands often reduce vanadium pentafluoride to the quadrivalent state, so that simple adducts of the type $VF_5.L$ or $VF_5.2L$ are not known. Indeed recent research clearly shows[225] that the pentafluoride is a strong oxidizing and fluorinating agent, being more active in this regard than UF_6, MoF_6, WF_6, NbF_5 and TaF_5. The reactions between the pentafluoride and covalent chlorides are complex, a large number of different products being formed. It reacts with XeF_6 and $XeOF_4$ to form the compounds $(XeF_6)_2.VF_5$ and $(XeOF_4)_2.VF_5$ the structures of which are unknown[226].

The best defined complexes of vanadium(V) are the hexafluorovanadates M^IVF_6 (M = K or Ag) and $Ba(VF_6)_2$. These may be prepared by the action of bromine trifluoride on a mixture of the appropriate chloride and vanadium trichloride[227]. The salts are white, with the exception of the orange-red silver salt, and diamagnetic. They are very readily hydrolysed, fuming in moist air, and are thus less stable than the corresponding niobium and tantalum salts. Moreover, there is apparently no tendency for the VF_6^- ion to add further fluoride ions to become seven- or eight-coordinate in the manner established for niobium and tantalum (cf. the NbF_7^{2-} and TaF_8^{3-} ions).

A nitrosyl salt $NO^+VF_6^-$ is also known; it can be prepared by the action of NOCl in excess and BrF_3 on vanadium pentoxide[228].

224 C. K. Jorgensen, *Absorption Spectra and Chemical Bonding in Complexes*, Pergamon, Oxford (1962).
225 J. H. Canterford and T. A. O'Donnell, *Inorg. Chem.* **5** (1966) 1442; *ibid.* **6** (1967) 541.
226 G. J. Moody and H. Selig, *J. Inorg. Nucl. Chem.* **28** (1966) 2429.
227 H. J. Emeléus and V. Gutmann, *J. Chem. Soc.* (1949) 2979.
228 A. G. Sharpe and A. A. Woolf, *J. Chem. Soc.* (1951) 798.

35. THE CHEMISTRY OF NIOBIUM AND TANTALUM*

D. Brown
A.E.R.E., Harwell

1. THE ELEMENTS

1.1. INTRODUCTION AND HISTORICAL

Niobium and tantalum, the heaviest elements of Group VA, are closely associated in nature and although they were discovered almost simultaneously the establishment of their individual identities was hindered by their close chemical similarities. In 1801 an English chemist, Charles Hatchett, isolated a white oxide[1] of a new element from a black mineral which had been presented to the British Museum in 1753. He named the new element columbium in honour of the country of its origin and the name "columbite" was subsequently applied to the black mineral. In the following year the Swedish scientist Ekeberg[2], who was studying minerals obtained from Kimito (Finland) and Ytterby (Sweden), isolated an element which he named tantalum, from the Greek *tantalos* because of the difficulty of dissolving the oxide of the new element.

Following these discoveries the identities of columbium and tantalum became confused and it was claimed by Wollaston[3] that they were, in fact, one and the same element. This view persisted until Heinrich Rose made a thorough study of a variety of "columbites" and "tantalites". Rose isolated two elements, one tantalum, and another which he named niobium from "Niobe" the daughter of "Tantalos"[4]. He also believed[5] he had isolated another element (pelopium) but later[6] decided that this was in fact niobium. The subsequent work of Bloomstrand[7], Marignac[8] and Deville and Troost[9] was instrumental in resolving the separate identities of niobium and tantalum. The first reliable values for the atomic weights of these elements were obtained by Marignac[8] who also devised the first commercially successful procedure for their separation.

The name niobium has been used almost exclusively in Europe but, despite the recommendations of the Commission on Inorganic Nomenclature and the Atomic Weights Commission, made at the Amsterdam Conference of the International Union of Pure and Applied Chemistry, that this name be accepted, the name columbium (symbol Cb) is still used, particularly by American metallurgists and mineralogists and American Industry.

* Literature covered to April 1969.

1 C. Hatchett, *Phil. Trans.* (1802) 49.
2 A. G. Ekeberg, *Ann. Chim.* **43** (1802) 276.
3 W. H. Wollaston, *Phil. Trans.* (1809) 246.
4 H. Rose, *Pogg. Ann.* **63** (1844) 317.
5 H. Rose, *Pogg. Ann.* **69** (1846) 115.
6 C. Rose, *Pogg. Ann.* **90** (1853) 456.
7 C. W. Bloomstrand, *J. Prakt. Chem.* **97** (1866) 37.
8 M. C. Marignac, *Ann. Chim. Phys.* **8**, 4 (1866) 5; ibid. **9**, 4 (1866) 249.
9 H. St. C. Deville and L. Troost, *Compt. Rend.* **60** (1865) 1221; ibid. **64** (1867) 294.

1.2. OCCURRENCE AND DISTRIBUTION

Niobium and tantalum do not occur in the free state and, unlike many other elements, they do not occur as sulphides. They are usually found together in combination with oxygen and one or more other metals. Although estimates vary greatly niobium is approximately as abundant as nickel and latest figures indicate that it is approximately 20 times more abundant than tantalum. The geochemistry of niobium and tantalum is covered in a recent review article.[10]

The columbite–tantalite series of minerals, of general formula $(Fe, Mn)(Nb, Ta)_2O_6$, have traditionally been considered as the most important sources of the elements. They are found in granite pegmatites and in residual or alluvial deposits derived from such rock. A desirable columbite is mined in Nigeria (Nb:Ta, 10:1) and important tantalo-columbites of intermediate composition are found in the Belgian Congo, whilst tantalite is mainly produced in Brazil and Australia. The tantalite concentrates, however, are relatively scarce and much of the tantalum produced is derived from concentrates containing greater amounts of niobium. Reserves of columbite are insufficient to support large-scale production of niobium and tantalum and the minerals pyrochlore and euxonite have gradually assumed industrial importance. Microlite, the tantalum-rich end of the pyrochlore–microlite group, contains up to 80% Ta_2O_5 but it is relatively rare and little appears to be known about reserves. Pyrochlore usually occurs in pegmatite derived from alkalic rock; reserves are found, for example, in Norway, Canada and Brazil. Large quantities of niobium- and tantalum-bearing concentrates are produced as by-products of tin mining operations in the Belgian Congo, Nigeria and Malaya. Details of mineral reserves, availability and composition and the demand for niobium and tantalum are discussed elsewhere[10–12]. The compositions of some important niobium- and tantalum-containing minerals are shown in Table 1; more extensive lists are available in various reviews[10–12].

TABLE 1. COMPOSITION OF SOME NIOBIUM AND TANTALUM MINERALS[a]

Mineral	General formula	Nb_2O_5 $+Ta_2O_5$ (%)	Nb_2O_5 (%)	Ta_2O_5 (%)
Columbite	$(Fe, Mn)(Nb, Ta)_2O_6$	75–81	47–78	Tr–34*
Tantalite	$(Fe, Mn)(Ta, Nb)_2O_6$	81–86	2–27	53–84
Tapiolite	$FeTa_2O_6$	80–85	1–11	74–83
Pyrochlore	$NaCaNb_2O_6F$	38–73	26–73	0·2–22·0
Microlite	$(Na, Ca)_2Ta_2O_6(O, OH, F)$	58–80	3–30	33–77
Euxenite	$(Y, Ca, Ce, U, Th)(Nb, Ta, Ti)_2O_6$	28–31	22–29	1–5

* Tr = trace.
[a] J. M. Warde, V. A. Mee and H. R. Spedden, Chapter 1 in *Columbium and Tantalum*, Ed. F. T. Sisco and E. Epremian, Wiley and Sons, New York (1963).

[10] R. L. Parker and M. Fleischer, U.S. Geol. Surv. Prof. Pap., 1968, No. 612.
[11] J. M. Warde, V. A. Mee and H. R. Spedden, Chapter 1 in *Columbium and Tantalum*, Ed. F. T. Sisco and E. Epremian, Wiley and Sons, New York (1963).
[12] G. L. Miller, *Tantalum and Niobium*, Butterworths, London (1959).

1.3. EXTRACTION PROCEDURES AND INDUSTRIAL USES

The association of niobium and tantalum in nature, coupled with their chemical similarity, requires that any extraction procedure for one involves the separation from the other. These factors have been the major drawbacks in the development of efficient processes for the commercial production of niobium and tantalum because the ultimate aim has been to achieve high yields of both elements from a given source. The various processes used for the isolation of these elements from ores are described fully in earlier reviews[12–17] and only brief accounts are given here. In such processes the ore-opening stage generally involves alkali fusion, digestion with acids (usually hydrofluoric acid) or chlorination, the treatment being applied either directly to the ore or following pretreatment such as roasting, conversion to crude carbides or reduction to a ferro-alloy.

Until recently the only industrial method for separating niobium and tantalum involved the crystallization of K_2TaF_7 from dilute hydrofluoric acid solution (the Marignac process[8]). The niobium complex formed under these conditions, $K_2NbOF_5 . 2H_2O$, is appreciably more soluble than K_2TaF_7. This process, which has been used since 1866, yields tantalum free from niobium and traces of tin, tungsten, titanium, etc., but, because the impurities accumulate in the mother liquor, the production of pure niobium is more difficult. This is generally achieved by precipitating sodium niobate, converting this to niobic acid by treatment with hydrochloric acid and then preparing potassium niobate which can be purified by recrystallization.

Such procedures are rapidly being superseded by industrial-scale liquid–liquid extraction methods. Suitable aqueous phases are hydrofluoric acid, hydrofluoric acid–hydrochloric acid mixtures and hydrofluoric acid–sulphuric acid mixtures. Milner and Wood[18] first observed that tantalum could be extracted from hydrofluoric acid by ketones and the results of later extensive studies have shown[19–27] that excellent separations can be achieved in either of two ways. In one[24], tantalum is extracted into methyl-isobutyl ketone from dilute hydrofluoric acid following which the acidity is increased by the addition of either hydrochloric acid or sulphuric acid and the niobium then extracted into a fresh batch of the same solvent. The elements are both recovered from the organic extract by stripping with water. Alternatively[26], following simultaneous extraction from 20% hydrofluoric acid solution the elements can be preferentially stripped from the methyl isobutylketone using water since the acid back extracted by the first water washes is sufficient to ensure that

[13] R. A. Gustison and R. L. Pilloton, ref. 11, chapter 2.

[14] E. M. Sherwood, Chapter 9 in *The Rare Metals Handbook*, 2nd ed., Ed. C. A. Hampel, Reinhold Pub. Co., Chapman & Hall, London (1961).

[15] C. A. Hampel, ref. 14, chapter 25.

[16] R. W. Moshier, *Analytical Chemistry of Niobium and Tantalum*, Pergamon, Oxford (1964).

[17] S. Kallman, Chapter 3 in *Treatise on Analytical Chemistry*, Part II, Vol. 6, Ed. I. M. Kolthoff and P. J. Elving, Interscience, New York (1964).

[18] G. W. C. Milner and A. J. Wood, Report AERE-C/R 895 (1952).

[19] K. B. Higbie and J. R. Werning, Bur. Mines Report of Invest. 5239 (1956).

[20] R. A. Foos and H. A. Wilhelm, U.S.A.E.C. Reports ISC-458 and ISC-694 (1954).

[21] J. R. Werning and K. B. Higbie, *Ind. Eng. Chem.* 46 (1954) 2491.

[22] J. R. Werning, K. B. Higbie, J. T. Grace, B. F. Speece and H. L. Gilbert, *Ind. Eng. Chem.* **46** (1954) 644.

[23] I. O. Winsch and L. Burris (Jr.), *Chem. Eng. Prog.* **53** (1957) 273.

[24] C. H. Chilton, *Chem. Eng.* **65** (1958) 104.

[25] J. L. Tewes and S. L. Mays, in *Technology of Columbium*, p. 36, Ed. B. W. Ganser and E. M. Sherwood, Wiley and Sons, New York (1958).

[26] E. L. Koerner (Jr.), M. Smutz and H. A. Wilhelm, *Chem. Eng. Prog.* **54** (1958) 63.

[27] P. C. Stevenson and H. G. Hicks, *Anal. Chem.* **25** (1953) 1517.

niobium remains in the organic phase. The advantages of a hydrofluoric acid–sulphuric acid–methyl isobutylketone system have been described[25] and a more efficient process based on the back extraction of niobium using sulphuric acid has been reported[28].

Other potentially useful isolation and separation procedures involving, for example, distillation of fluorides or chlorides and preferential hydrogen reduction of $NbCl_5$ do not yet appear to have been exploited commercially. Niobium metal is said to have been prepared first by Bloomstrand[29], who reduced the pentachloride with hydrogen in 1866. Later the pentoxide was reduced with carbon by Moissan[30] and with aluminium by Goldschmidt and Cantin[31]. Although Rose[32] prepared tantalum powder by reducing Na_2TaF_7 with metallic sodium the first pure samples of both niobium and tantalum were obtained by Von Bolton[33] who reduced the fluorometallates(V) with sodium. Tantalum was, in fact, used as the first metallic filament for incandescent lamps.

Following conversion of the separated elements to the pentoxides the appropriate metals are now obtained[12, 14, 15, 17, 34] by reduction with either carbon or the metal carbide. Alternatively, sodium reduction of niobium pentachloride or potassium heptafluorometallates(V), or electrolysis of K_2TaF_7 mixed with Ta_2O_5 can be employed. The last process, which is extensively used for tantalum production[12, 35], is less satisfactory when applied to niobium owing to the greater ease of reduction of niobium and the formation of stable oxygenated niobium species when fluoroniobates(V) and niobium pentoxide are used. Niobium is best prepared[36] by reacting the monocarbide, prepared from pentoxide, with the stoichiometric amount of the pentoxide for the reaction,

$$Nb_2O_5 + 5NbC \xrightarrow{1600\text{–}2500^\circ} 7Nb + 5CO$$

Consolidation and purification of the products from these reactions are achieved by sintering, vacuum or inert atmosphere arc melting, electron beam melting or iodide refining.

Niobium and tantalum are widely used in industry[12, 14, 15, 17, 37]. Although pure niobium finds relatively few applications it is added to austenitic chromium nickel type stainless steels in the form of ferroniobium or ferroniobium–tantalum as a stabilizer to prevent carbide precipitation. It inhibits intergranular corrosion and improves strength at high temperature; in 4 to 6% chrome steels the presence of niobium reduces air hardening, improves impact strength, reduces creep, shortens annealing time and improves resistance to oxidation. Numerous high-temperature alloys containing niobium (0·44 to 4%) are used in components for aircraft jet engines and turbine wheels. Because of its strength at high temperatures, its moderately low neutron capture cross-section and its resistance to liquid metals, niobium is a potential constituent of nuclear fuels and fuel cladding materials. Niobium-bearing electrodes are used in welding and a minor use is found as a "gas-getter" in electronic tubes.

[28] D. J. Soissan, J. L. McLafferty and J. A. Pierret, *Ind. Eng. Chem.* **53** (1961) 861.
[29] C. W. Bloomstrand, *J. Prakt. Chem.* **99** (1866) 40.
[30] H. Moissan, *Compt. Rend.* **133** (1901) 20.
[31] H. Goldschmidt and C. Vantin, *J. Soc. Chem. Ind.* **19** (1898) 543.
[32] H. Rose, *Pogg. Ann.* **99** (1856) 40.
[33] W. von Bolton, *Z. Elektrochem.* **13** (1907) 145.
[34] J. H. Downing and D. J. Hansen, ref. 11, chapter 3.
[35] R. Kieffer and H. Braun, *Vandin, Niob, Tantal*, p. 68, Springer-Verlag, Berlin and Heidelberg (1963).
[36] K. C. Li, *J. Metals*, **12** (1960) 485.
[37] F. H. Buttner, F. E. Bacon and R. K. Bancroft, ref. 11, chapter 13.

Tantalum is extremely stable to most acids and this corrosion resistance coupled with its high strength, weldability, ductility and high heat transfer makes it a suitable construction material for a large range of chemical equipment. It forms an anodic oxide film of unusual stability, with a dielectric constant much greater than that of aluminium oxide, and this property is utilized in electrolytic rectifiers for operating railway signals and in the

TABLE 2. ISOTOPES OF NIOBIUM AND TANTALUM[a]

Isotope	Half-life	Type of decay	Principal means of production
Nb^{89}	1·9 hr	β^+	C^{12} on Br; $Y^{89}(\alpha, 4n)$
Nb^{89m}	42 m	β^+	C^{12} on Br
Nb^{90}	14·6 hr	β^+, E.C.	$Zr^{90}(p, n)$; $Zr^{90}(d, 2n)$
Nb^{90m}	24 sec	I.T.	Daughter Mo^{90}
Nb^{91}	long	E.C.	$Zr^{90}(d, n)$
Nb^{91m}	64 d	I.T. (97%); E.C. (3%)	$Y^{89}(\alpha, 2n)$; $Zr^{90}(d, n)$
Nb^{92}	>350 y or <1 hr		
Nb^{92m1}	10·16 d	E.C. (99+%); β^+ (0·06%)	$Y^{89}(\alpha, n)$
Nb^{92m2}	13 hr	E.C.	Protons on Nb^{93}
Nb^{93}	stable	—	Natural
Nb^{93m}	13·6 y	I.T.	Daughter Zr^{93}
Nb^{94}	2×10^4 y	β^-	$Nb^{93}(n, \gamma)$
Nb^{94m}	6·29 m	I.T. (99+%); β^- (0·2%)	$Nb^{93}(n, \gamma)$
Nb^{95}	35·0 d	β^-	Daughter Zr^{95}
Nb^{95m}	90 hr	I.T.	Daughter Zr^{95}
Nb^{96}	23·35 hr	β^-	$Zr^{96}(p, n)$; $Mo^{98}(d, \alpha)$
Nb^{97}	72 m	β^-	Descendant of Zr^{97}
Nb^{97m}	1·0 m	I.T.	Daughter Zr^{97}
Nb^{98}	51 m	β^-	$Mo^{98}(n, p)$
Nb^{98m}	<2 m	β^-	Fission, daughter Zr^{98}
Nb^{99}	2·4 m	β^-	Fission, $Mo^{100}(\gamma, p)$
Nb^{100}	3·0 m	[β^-]	Fission
Nb^{100m}	11 m	β^-	$Mo^{100}(n, p)$
Nb^{101}	1·0 m genet	[β^-]	Fission
Ta^{172}	44 m	β^+; E.C.	Protons on Hf^{176}
Ta^{173}	3·7 hr	E.C.; β^+	$Ho^{165}(N^{14}, 6n)W^{173}$(E.C.); protons on Ta^{181}
Ta^{174}	1·2 hr	E.C.; β^+	$Ho^{165}(N^{14}, 5n)W^{174}$(E.C.); protons on Ta^{181}
Ta^{175}	10·5 hr	E.C.	$Lu^{175}(\alpha, 4n)$; $Hf^{176}(p, 2n)$
Ta^{176}	8·0 hr	E.C.	$Lu^{175}(\alpha, 3n)$; $Hf^{176}(p, n)$
Ta^{177}	56·6 hr	E.C.	$Lu^{175}(\alpha, 2n)$; protons on Hf
Ta^{178}	9·35 m	E.C. (99%); β^+ (1%)	Daughter of W^{178}
Ta^{178m}	2·1 hr	E.C.	$Lu^{175}(\alpha, n)$; protons on Hf
Ta^{179}	~600 d	E.C.	$Lu^{176}(\alpha, n)$; protons on Ta^{181}
Ta^{180}	$>10^{12}$ y	—	Natural (0·0123%)
Ta^{180m}	8·15 hr	E.C. (87%); β^- (13%)	$Hf^{180}(d, 2n)$; $Ta^{181}(n, 2n)$
Ta^{181}	stable	—	Natural (99·9877%)
Ta^{181m}	0·33 sec	I.T.	Neutrons on Ta
Ta^{182}	115·1 d	β^-	$Ta^{181}(n, \gamma)$
Ta^{182m}	16·5 m	I.T.	$Ta^{181}(n, \gamma)$
Ta^{183}	5·0 d	β^-	$Ta^{181}(n, \gamma)$; $Ta^{182}(n, \gamma)$
Ta^{184}	8·7 hr	β^-	$W^{186}(d, \alpha)$; $W^{184}(n, p)$
Ta^{185}	50 m	β^-	$W^{186}(\gamma, p)$; $W^{186}(n, pn)$
Ta^{186}	10·5 m	β^-	$W^{186}(n, p)$

[a] C. M. Lederer, J. M. Hollander and I. Perlman, *Table of Isotopes*, 6th ed., Wiley and Sons, New York (1967).

construction of capacitors. The advantages of tantalum capacitors are greater life, wider temperature range, higher capacitance-to-size ratio and a better idle-life. Tantalum is used in the construction of special-purpose vacuum tubes, and, owing to its inertness to body tissue and fluids, in surgery as suture wire and cranial repair plates. It also finds use as electroplating anode holders and spinnerettes for artificial fibres.

1.4. NUCLEAR PROPERTIES

Niobium-93, tantalum-180 and tantalum-181 are the only naturally occurring isotopes. The numerous radioactive niobium and tantalum isotopes known at present[38] are listed in Table 2 together with their modes of decay and methods of preparation. The most useful of these for radioactive tracer studies are niobium-95 and tantalum-182; the latter is also of considerable analytical importance for determinations using neutron activation analysis[39].

1.5. PHYSICAL AND CHEMICAL PROPERTIES OF THE METALS

A summary of the more important physical properties of niobium and tantalum is provided in Table 3. These properties and the mechanical and chemical properties are discussed in detail elsewhere[12, 14, 15, 40]. Niobium and tantalum, which are not polymorphic, crystallize with body-centred cubic symmetry, thus resembling the larger alkali metals. As a consequence of the lanthanide contraction the ionic radii of these elements are virtually identical and this is reflected in the close resemblance of many of their chemical properties.

Niobium is a far more reactive metal than tantalum and when being processed even at moderate temperatures it must be protected by a vacuum or inert atmosphere. Protective oxide films form on both metals but niobium is dissolved readily in hydrofluoric acid and aqueous alkali solutions. The corrosion resistance of niobium is inferior to that of tantalum and it is attacked by most reagents to which tantalum is inert. Attack by acids releases sufficient hydrogen to embrittle niobium. The corrosion resistance of tantalum has resulted in its extensive use for the construction of chemical equipment. Details of the corrosion resistance of both metals are provided and discussed in earlier reviews[12, 14, 15, 41].

Fuming nitric acid and hot concentrated hydrochloric, sulphuric and phosphoric acids (but not concentrated nitric acid) attack niobium whereas tantalum is inert when exposed to these acids up to moderately high temperatures. Both metals are unaffected by these acids and a host of other chemicals at room temperature but are attacked by fluoride in acid media and by fuming sulphuric acid.

Tantalum is attacked by concentrated alkaline solutions at room temperature and dissolves in molten alkalis. However, unlike niobium, it is quite resistant to dilute solutions. Both metals are very resistant to attack by liquid metals such as lithium, sodium, potassium, sodium–potassium alloy, bismuth, lead, mercury and gallium provided these liquid metals contain no oxygen. In general niobium is attacked at lower temperatures than tantalum.

[38] C. M. Lederer, J. Hollander and I. Perlman, *Table of Isotopes*, 6th ed., Wiley and Sons (1967).
[39] E. N. Jenkins and A. A. Smales, *Quart. Rev.* **10** (1956) 83.
[40] R. W. Fountain and C. R. McKinsey, ref. 11, chapter 7.
[41] M. Stern and R. C. Bishop, ref. 11, chapter 8.

TABLE 3. SOME PHYSICAL PROPERTIES OF NIOBIUM AND TANTALUM*

Property (units)	Niobium	Tantalum
Atomic number[a, b]	41	73
Atomic weight[a, b]	92·91	180·95
Atomic volume (cc g/atom)[a, b]	10·83	10·90
Crystal structure[a, b]	bcc A_2	bcc A_2
Lattice constant (Å)[a, b]	3·294	3·300
Coordination number[a, b]	8	8
Density at 20°C (g cm^{-3})[a, b]	8·57	16·60
Valence electron configuration[b]	$4d^4 5s^1$	$5d^4 6s^1$
Atomic radius (Å)[a]	1·47	1·47
Thermal neutron capture cross-section (barns/atom)[b]	1·15±0·05	21·3±1·0
Melting point (°C)[a]	2468±10	2980±50
Boiling point (°C)[a]	4927	5429
Heat of combustion (cal/g)[b]	2379	1346[a]
Latent heat of fusion (cal/mol)[b]	6400	7500
Latent heat of vaporization (cal/g)[b]	166,500	180,000
Heat of sublimation at 0°K[b] (cal/g atom)	171,800	185,500
Heat of sublimation at 15°K (cal/g atom)	172,530	—
Heat capacity (cal/°C/g atom)[a]	$5{\cdot}66+0{\cdot}96T$ (298–1900°K)	$5{\cdot}80+0{\cdot}71T+0{\cdot}05\times10^{-6}T^2$ (298–2300°K)
Specific heat (cal/g/°C)[a, b]	0·064	0·0340
Entropy (cal/°C/g atom)[a, b]	8·73	9·90
Enthalpy at 15°C (cal/g atom)[b]	1264	
Vapour pressure ($\log p_{mm}$)[a]	$-\frac{37{,}650}{T}+8{\cdot}94+0{\cdot}715\log T-0{\cdot}166T$ (298°—m.p.)	$-\frac{40{,}800}{T}+10{\cdot}29$ (290°—m.p.)
Thermal conductivity[b] (cal/cm^2/°C/cm)	0·125 at 0°C 0·135 at 200°C 0·156 at 600°C	0·130 at 20°C 0·1625 at 568°C 0·1715 at 828°C
Coefficient of linear thermal expansion (cm/cm/°C)[b]	$7{\cdot}88\times10^{-6}$ (0–1000°C)	$8{\cdot}0\times10^{-6}$ (20–1500°C)
Electrical resistivity (temperature coefficient of)[b]	$3{\cdot}96\times10^{-3}$ (0–600°C)	$3{\cdot}0\times10^{-3}$ (0–1000°C)
Transition temperature (°K)[b]	9·25	4·38
Magnetic susceptibility (c.g.s. units)[b]	$2{\cdot}28\times10^{-6}$ (at 25°C)	$0{\cdot}849\times10^{-6}$ (at 25°C) $0{\cdot}685\times10^{-6}$ (at 1870°C)
Work function (eV)[b]	4·0	4·12
Positive ion emission (eV)[b]	5·52	10·0
Electrochemical equivalent (mg/coulomb)[b]	0·1926	0·3749
Refractive index[c]	1·80	2·05
Standard electrode potential for M/M^{5+}[c]	0·96	1·12
Ionization potential (V)[c]	6·67	7·3±0·3

[a] *Metals Reference Book*, Ed. C. J. Smithells, Vols. I and II, Butterworths, London (1967).
[b] *Rare Metals Handbook*, Ed. C. A. Hampel, 2nd ed., pp. 149 and 469, Reinhold, London (1961).
[c] R. W. Fountain and C. R. McKinsey, Chapter 7 in *Columbium and Tantalum*, Ed. F. T. Sisco and E. Epremian, Wiley and Sons, New York (1963).

* See also Volume I of *Thermophysical Properties of High Temperature Solid Materials*, Ed. Y. S. Touloukian, MacMillan, New York (1967).

Niobium is oxidized by water vapour at about 250° with the formation of a surface film of NbO; between 950° and 1300°C tantalum is oxidized to Ta_2O_5. The kinetics of the reactions of the elements with oxygen, nitrogen and carbon have been extensively investigated and the results have been reviewed[12, 14, 15, 42]. The various compounds formed with these and other elements are discussed later.

1.6. PURIFICATION AND ANALYTICAL CHEMISTRY

Numerous methods are available for the separation of niobium and tantalum from each other and from a wide variety of other elements. Such procedures, apart from certain solvent extraction systems, find little industrial application, but they are frequently employed during analysis. The presence of other elements may cause serious difficulties during analytical separations involving niobium and tantalum. As a result of the abnormal behaviour observed[17] these difficulties are often referred to as "loss of individuality". No attempt will be made to provide a complete coverage of the wide range of purification and analytical methods available and the reader is referred to other review articles for detailed discussions[16, 17, 43–47].

A wide range of reagents can be used, under strictly specified conditions, in precipitation reactions for the separation of niobium and tantalum from a variety of impurities, or from each other. These include, for example[16, 17], selenous acid, cupferron, tannin, phenylarsonic acid, pyrogallol, hexamethylenetetramine, cinnamylhydroxamic acid, 8-hydroxyquinoline, N-benzoyl,N-phenylhydroxylamine and n-propyl arsenic acid. Such methods, however, are rapidly being replaced by anion exchange and solvent extraction procedures, particularly since the separation of niobium from tantalum is greatly improved and simplified. Niobium(V) and tantalum(V) are usually present in solution as complex anions and, therefore, cation exchange finds little application in separation procedures[17, 45].

Although anion exchange in hydrochloric acid media can be used it is limited to tracer amounts of niobium and tantalum owing to the hydrolysis which occurs in such solutions. Excellent separations from other elements can be achieved in hydrofluoric acid but the separation of niobium from tantalum requires care[49]. The most useful media for the purification of niobium and tantalum by anion exchange are undoubtedly mixtures of hydrofluoric acid with either hydrochloric acid, ammonium fluoride or oxalic acid. Compilations of the anion exchange behaviour of many elements in hydrofluoric acid[48], hydrofluoric acid–hydrochloric acid[49–51] and sulphuric acid[52] are available and many useful separations are discussed in detail by Kallmann[17] and by Korkisch[45] together with separations based on cellulose and paper chromatography.

42 F. E. Bacon and P. M. Moanfeldt, ref. 11, chapter 9.
43 H. Green, *Metallurgia*, **70** (1964) 299.
44 E. P. Steinberg, *The Radiochemistry of Niobium and Tantalum*, U.S. Report NAS-NS 3039 (1961).
45 J. Korkisch, *Modern Methods for the Separation of Rarer Metal Ions*, p. 455, Pergamon Press, Oxford (1969).
46 W. R. Schoeller and A. R. Powell, *The Analysis of Minerals and Ores of the Rarer Elements*, 3rd ed., Griffin, London (1955).
47 F. Cuttitta, U.S. Geol. Survey Bull. 1029-A (1957).
48 M. K. Nikitin, *Dokl. Akad. Nauk. SSSR*, **148** (1963) 595.
49 J. L. Hague, E. D. Brown and H. A. Bright, *J. Res. Natl. Bur. Stds.* **53** (1954) 261.
50 K. A. Kraus and F. Nelson, ASTM Spec. Tech. Publ. No. 195 (1958).
51 L. Wish, *Anal. Chem.* **31** (1959) 327.
52 L. R. Bunney, N. E. Ballou, J. Pascual and S. Foti, *Anal. Chem.* **31** (1959) 324.

Industrial processes for the separation and purification of niobium and tantalum by solvent extraction have already been mentioned. Separations based on the hydrofluoric acid–sulphuric acid, hydrofluoric acid–hydrochloric acid and hydrofluoric acid–nitric acid systems can be achieved[17, 45] using ketones (particularly methyl isobutylketone), carbinols, cyclohexanone, tributyl phosphate, amines or mesityl oxide as extractants. The extraction of chelates formed with, for example, oxine, cupferron and 5,7-dichloro-8-quinolinol and the use of the di-n-butyl ester of phosphoric acid have also been studied[17]. As with ion exchange the use of hydrochloric acid alone is less satisfactory owing to the hydrolysis of niobium and tantalum at macro-concentrations.

The methods available for the determination of niobium and tantalum are discussed at length in the excellent review by Kallmann[17] who also deals with the determination of metallic impurities in the elements. A recent comprehensive review deals with the rôle of chelating agents in the analytical chemistry of niobium and tantalum[53]. Following purification (if necessary) by one of the methods discussed above, niobium and tantalum can be determined gravimetrically, spectrophotometrically, volumetrically or by a variety of other instrumental techniques.

Gravimetric determination generally involves precipitation with reagents such as cupferron, tannic acid, phenylarsonic acid or precipitation by hydrolysis in ammoniacal solution followed by ignition to the pentoxide. Niobium can be determined spectrophotometrically using a variety of reagents of which the most widely employed are the thiocyanate complex in ether, the peroxide or hydroquinone complex in sulphuric acid, and the oxinate complex in chloroform. Fewer spectrophotometric methods are available for tantalum; these include the use of complexes with pyrogallol (acid solution), hydroquinone, gallic acid and hydrogen peroxide. The chief advantage of the colorimetric method lies in its application to solutions containing only small amounts of the elements. Although niobium can be determined titrimetrically following reduction using either a Jones reductor or cadmium, and also polarographically, such methods are not widely employed. The titrimetric method was exhaustively studied[55] several years ago; more recently an improved procedure involving hydrofluoric acid solution has been reported[56].

Both elements can be determined by the usual techniques of arc or spark spectrographic analysis. For niobium the most sensitive lines are at $\lambda = 4058{\cdot}94$ and $5079{\cdot}73$ whilst for tantalum the most sensitive ones are at $\lambda = 2653{\cdot}27$ and $2714{\cdot}67$ [17, 57]. It is possible to detect about 0·05% niobium and 0·5% tantalum in steel using the spark technique. These limits are considerably lower, 0·001% and 0·03% respectively, when X-ray fluorescence methods are used. In view of the problems associated with the intermetallic diffusion of niobium with the components of stainless steel recent interest has centred on the determination of niobium using an electron probe microanalyser[58].

Niobium-95 ($t_{\frac{1}{2}} = 35{\cdot}0$ d) and tantalum-182 ($t_{\frac{1}{2}} = 115{\cdot}1$ d), both of which are β^- emitters, can be determined by routine radiochemical procedures and have been widely used for monitoring various separation procedures and in conjunction with an isotope

53 E. Lassner and R. Püschel, *Chelates in Anal. Chem.*, Vol. 2, p. 213, Ed. H. A. Flaschka and A. J. Barnard, Jr. (1969).
54 C. Djordjevic and D. Sevdi, *Croat. Chem. Acta*, **39** (1963) 155.
55 H. B. Knowles and G. E. F. Lundell, *J. Res. Natl. Bur. Stand.* **42** (1949) 405.
56 J. B. Headridge and M. S. Taylor, *Analyst*, **87** (1962) 43.
57 W. F. Meggers, C. H. Corliss and B. F. Scribner, Natl. Bur. Stds. Monograph 32 (1961).
58 R. E. Seebold and L. S. Birks, *J. Nucl. Materials*, **3** (1961) 260, 249.

dilution technique for the determination of the elements in granite[59]. It should be noted that dilution techniques require a sufficiently high concentration of complexing agent to prevent hydrolysis with subsequent radiocolloid formation and to ensure complete exchange.

Neutron activation analysis can be applied to tantalum compounds and it has been estimated that by measurement of the β-activity of the resulting tantalum-182 as little as 10^{-10} g of the element can be determined under ideal conditions[39]. An analysis involving shorter irradiations is possible using the isotope tantalum-182m ($t_{\frac{1}{2}}$ = 18·5 m) but the sensitivity is only 4×10^{-8} g. This technique is of very much less value for the determination of niobium owing to the low neutron capture cross-section of niobium-93 (1·1 barns) and the short half-life of the resulting niobium-94m (6·29 m).

2. COMPOUNDS OF THE ELEMENTS

2.1. INTRODUCTION

The chemistries of niobium and tantalum are very similar to each other but are significantly different from that of their Group VA congener, vanadium, and bear only a slight resemblance to those of the Group VB elements. The chemical and crystallographic similarities between niobium and tantalum arise from their identical atomic (1·47 Å) and ionic (0·69 Å for M^{5+}) radii (consequences of the lanthanide contraction) which are appreciably larger than the corresponding values for vanadium, 1·34 Å and 0·59 Å, respectively. One consequence of the size difference is that the lower oxidation states of niobium and tantalum are of relatively minor importance, a feature which is, of course, characteristic of the other vertical *d*-transition element groups. Thus, apart from the halide chemistry, the vast majority of characterized niobium and tantalum compounds are pentavalent and although compounds with formal oxidation states of +4, +3, +2, +1, 0 and −1 have been reported these oxidation states, in particular the first three, are less well characterized than the corresponding ones for vanadium. In contrast to vanadium, for which the species VO_3^+ and, particularly, VO^{2+} play important roles, niobium and tantalum have virtually no cation chemistry and even $NbOCl_3$ exists as an oxygen-bridge polymer in the solid state whereas $VOCl_3$ contains the $V{=}O^{3+}$ group. Differences are also observed in the chemistry of the oxides, for example, Nb_2O_5 and Ta_2O_5 are appreciably more basic than V_2O_5 which is amphoteric, and in the chemistry of the lower halides where metal–metal bonding is important in the niobium and tantalum tetrahalides and in their numerous integral valence state cluster compounds. At present no vanadium analogues are known for the last class of compound.

There are some differences between the chemistries of niobium and tantalum but these are not profound. Thus, it is well established that niobium(V) is more readily reduced than tantalum(V) and is less readily hydrolysed in aqueous hydrochloric acid solutions from which complexes containing the $NbOCl_5^-$ ion but not the $TaOCl_5^-$ ion may be isolated. In addition, there are differences in the structural chemistry of the respective pentoxides, in the stabilities of the pentavalent oxyhalides, and in the nature of the oxysulphates obtained from aqueous solution. These and other examples are dealt with in the succeeding sections. The first ionization potentials for niobium and tantalum are 6·7 and 7·3 eV, respectively (Table 3), and the values of $I_2 \rightarrow I_5$ inclusive for niobium are 14·32, 25·04, 38·3 and 50·0 eV, respectively.

[59] A. D. Savostin and I. P. Alimarin, *Vestn. Moskov. Univ.* **45** (1960) (*Anal. Abst.* **7** (1960) 4227).

It will be obvious from the succeeding sections that the halides are the most extensively studied compounds of niobium and tantalum. More recently interest has centred on the characterization of both tetravalent and pentavalent halide complexes with oxygen, sulphur and nitrogen donors, generally by reactions in non-aqueous solvents, and on the organometallic chemistry of the elements. Hydride chemistry, alloys and the numerous compounds formed with the elements of Groups IVB, VB and VIB have also received much attention, but relatively little is known concerning the cyanides, thiocyanates, nitrates, phosphates and sulphates. Borates, carbonates, nitrites, phosphites, and sulphites do not appear to have been characterized and examples of hypohalites, halites, halates and perhalates are also unknown.

2.2. ALLOYS

The formation of alloys between niobium or tantalum and other metals, particularly transition metals, has been extensively studied on account of the potential industrial application of the resulting materials. Both elements form binary alloys with many metals, exceptions being the alkali metals, alkaline earth metals, cadmium, copper, magnesium, lead, mercury, silver, yttrium and the lanthanides.

TABLE 4. SOME EXAMPLES OF ALLOYS FORMED BY NIOBIUM AND TANTALUM

Class M = niobium or tantalum A = alloying element	A	
	Niobium	Tantalum
M_3A	Re, Os, Ir, Pt, Au, Al, Ge, Sn	Sn, Co
M_2A	Al, Ge, Fe	Co
MA	Ir, Ni, Zn	Ni
MA_2	Be, Ge, Cr, Mn, Fe, Co, Zn	Ge, Cr, Mn, Fe
MA_3	Al, Ir, Ni, Rh, Fe	Al, Ir, Ni, Rh

With the high melting point metals such as β-titanium, hafnium, zirconium, vanadium, molybdenum and tungsten a continuous series of solid solutions are formed below the solidus. Some examples of the phases encountered in other binary systems are listed in Table 4. For detailed accounts of the preparation of these and other compounds, the reader is referred to available reviews[12, 60–64] which also contain information on crystallographic properties, phase diagrams, the alloying behaviour of niobium and tantalum in relationship to each other and to that of other elements, and the various ternary and quaternary systems which have been investigated. Compounds formed with the elements of Groups IV, V and VI, of which even those formed with non-metals are often referred to as alloys, are discussed in sections 2.10, 2.9 and 2.8, respectively.

[60] W. F. Sheely, ref. 11, chapter 10.

[61] A. E. Dwight, in *Columbium Metallurgy*, p. 383, Ed. D. L. Douglass and F. W. Kunz, Interscience, New York (1961).

[62] R. P. Elliott, *Constitution of Binary Alloys, First Supplement*, McGraw-Hill, New York (1965).

[63] W. B. Pearson, *Handbook of Lattice Spacings and Structures of Metals*, Vol. 2, Pergamon Press, London (1966).

[64] P. Feschotte, D. T. Livey and O. Von Goldbeck, *Atomic Energy Rev.* Special Issue, No. 2, 45 (1968).

The main interest lies in the fabrication of alloys of high strength, hardness and ductility at high temperatures, and of good strength–weight ratios. Despite the development of mechanically useful niobium-base alloys, however, it has not been possible to sufficiently improve the oxidation resistance of niobium by alloying to permit its widespread use as an elevated-temperature structural material in oxygen-containing environments. Although the corrosion resistance of tantalum alloys may be superior to that of tantalum in specific cases no alloy possesses the wide range of usefulness of tantalum itself in this respect. Important alloys are ferrocolumbium and ferrotantalum–columbium which are used to inhibit intergranular corrosion in austenitic steels and as alloying agents in certain special purpose steels. Because of their high corrosion resistance at moderately elevated temperatures and their low capture cross-section for thermal neutrons niobium-base alloys are of interest as claddings for fuel elements and as a matrix for fuel materials in nuclear reactors. Many niobium alloys become superconducting at easily attainable temperatures and since they remain superconducting in magnetic fields they are useful in the manufacture of superconducting high-field magnets. Details of other applications of niobium and tantalum alloys will be found in the review by Miller[12].

2.3. HYDRIDES

The niobium and tantalum hydrides are discussed fully in the recent authoritative review by Seigel and Libowitz[65] who deal with their preparation, phase diagrams, crystallographic properties and thermodynamic properties.

Both niobium and tantalum react rapidly with hydrogen at elevated temperatures to form a series of interstitial hydrides. The reactions are pressure-sensitive and an increase in pressure increases the rate of reaction. The process involves the dissociation of hydrogen molecules into atoms which then permeate through the crystal lattice of the metal and not along grain boundaries. The absorption is exothermic and the solubility of hydrogen is greatest at lower temperatures. Limiting compositions of $NbH_{0\cdot9}$ and $TaH_{0\cdot8}$ have been reported but, although it appears to be difficult to obtain phases of composition $MH_{1\cdot0}$, the niobium compound can apparently be prepared by reacting the components at 900° for 2 hr[66], and an unstable dihydride, NbH_2, is obtained[67] by cathodic hydrogenation in 6 M sulphuric acid solution. In a vacuum of 10^{-6} Torr this latter compound, which possesses the fluorite-type structure ($a_0 = 4\cdot563$ Å), rapidly decomposes to $NbH_{1\cdot0}$.

X-ray diffraction studies of the hydrides and deuterides of niobium and tantalum have shown that at room temperature a body-centred cubic structure (α-phase), based on that of the metals, persists up to compositions of about $NbH(D)_{0\cdot03}$ and $TaH(D)_{0\cdot2}$, respectively. At this temperature two-phase fields are found up to compositions of about $NbH(D)_{0\cdot75}$ and $TaH(D)_{0\cdot55}$, respectively, above which compositions the single β-phase is observed. This latter (β) phase has a distorted cubic structure in both systems and, for example, $NbH_{0\cdot9}$ is orthorhombic with $a_0 = 4\cdot84$, $b_0 = 4\cdot90$ and $c_0 = 3\cdot45$ Å whilst for $TaH_{0\cdot8}$ $a_0 = 4\cdot73$, $b_0 = 4\cdot78$ and $c_0 = 3\cdot43$ Å. At lower compositions in the tantalum hydride system the β-phase is tetragonal, for example at approximately $TaH_{0\cdot5}$ $a_0 = 3\cdot35$ and $c_0 = 3\cdot41$ Å. The two-phase field exists over a range of compositions and temperatures

65 B. Siegel and G. G. Libowitz, Chapter 12 in *Metal Hydrides*, Ed. W. M. Mueller, J. P. Blackledge and G. G. Libowitz, Academic Press, London (1968).
66 M. M. Antonova and G. V. Samsonov, *J. App. Chem. USSR*, **38** (1965) 2353.
67 G. Brauer and H. Müller, *J. Inorg. Nuclear Chem.* **17** (1961) 102.

with an upper limit of about 140° for the niobium–hydrogen system and $NbH_{0.7}$, for example[68], transforms from the distorted β-phase to the cubic modification at about 100°. In the niobium–hydrogen system a continuous solid solution is formed above 300°. This possesses the body-centred cubic structure, the lattice parameter and heat of solution increasing with hydrogen content. A similar solid solution phase (α) exists in the tantalum–hydrogen system at high temperatures and it is reported that a phase approximating to the composition $Ta_2H(D)$ precipitates out at lower temperatures. $Ta_2H(D)$ is of body-centred tetragonal symmetry with parameters close to that of the α-phase ($c/a = 1{\cdot}008$). At higher hydrogen concentrations this β-phase gradually distorts into a face-centred orthorhombic structure. Neutron diffraction studies[69] of Ta_2D have shown that below 33° the (β_1-phase) pseudo-cubic unit cell has $a_0 = 6{\cdot}74$ Å whilst at this temperature (β_2-phase) $a_0 = 3{\cdot}37$. The cell changes are interpreted on the basis of ordering of the deuterium atoms in the tetrahedral interstices in the low temperature form. Heat capacity data[70, 71] indicate that there are at least two, and possibly three, forms of β-Ta_2H. The residual entropy of this compound, $S_{0°K}$, is $-0{\cdot}13 \pm 0{\cdot}10$ e.u.[70].

The magnetic susceptibilities of niobium and tantalum hydrides are lower than those of the metals indicating the existence of spin interaction between the hydrogen and metal atoms. Thus, for $NbH_{0.86}$ and $TaH_{0.75}$ $\chi_m = 0{\cdot}70$ (291°K) and $0{\cdot}424$ (290°K) c.g.s. units, respectively.

The metal–hydrogen reactions are commercially important in the reprocessing of scrap metal, the preparation of powdered metals and in the production of the metals by reduction of the pentachloride, particularly since traces of absorbed hydrogen have a marked detrimental effect on the ductility of the metals. At 500° hydrogen is almost completely removed from $NbH_{0.9}$ and $TaH_{0.8}$ in 1 hr under a dynamic vacuum.

2.4. HALIDES, OXYHALIDES AND HALOGENO COMPLEXES

Introduction

The halides and oxyhalides of niobium and tantalum are probably the most extensively studied compounds formed by these elements and for comprehensive up-to-date reviews the excellent articles by Fairbrother[72, 73] and the one by Colton and Canterford[74] are recommended. The presently known anhydrous halides and oxyhalides are listed in Table 5. The fact that certain oxyhalides, for example certain of those of the types MO_2Br and $MOBr_2$, are unknown, is probably a reflection of lack of investigation rather than of their instability. These and other oxyhalides will probably be conveniently prepared by halide–antimony sesquioxide reactions similar to those recently reported for the preparation of certain actinide oxyhalides[75, 76]. The question of the existence of pure trifluorides is

[68] C. Wainwright, A. J. Cook and B. E. Hopkins, *J. Less-Common Metals*, 6 (1964) 362.
[69] W. E. Wallace, *J. Chem. Phys.* **35** (1961) 2156.
[70] W. G. Saba, W. E. Wallace, H. Sandmo and R. S. Craig, *J. Chem. Phys.* **35** (1961) 2148.
[71] W. E. Wallace, P. Kofstad and L. J. Hyvönen, *J. Pure and Appl. Chem.* **2** (1961) 281.
[72] F. Fairbrother, Halides of niobium and tantalum, in *Halogen Chemistry*, Ed. V. Gutmann, Vol. 1, p. 123, Academic Press, London (1966).
[73] F. Fairbrother, *The Chemistry of Niobium and Tantalum*, Elsevier, London (1967).
[74] J. H. Canterford and R. Colton, *Halides of the Second and Third Row Transition Metals*, Wiley and Sons, London (1968).
[75] D. Brown, J. F. Easey and P. J. Jones, *J. Chem. Soc.* (A) (1967) 1698.
[76] K. W. Bagnall, D. Brown and J. F. Easey, *J. Chem. Soc.* (A) (1968) 288.

TABLE 5. ANHYDROUS HALIDES AND OXYHALIDES OF NIOBIUM AND TANTALUM

Oxidation state	Fluoride*		Chloride		Bromide		Iodide	
V	NbF_5	TaF_5	$NbCl_5$	$TaCl_5$	$NbBr_5$	$TaBr_5$	NbI_5	TaI_5
	—	$TaOF_3$	$NbOCl_3$	$TaOCl_3$	$NbOBr_3$	$TaOBr_3$	$NbOI_3$	—
	—	—	—	$Ta_2O_3Cl_4$	—	—	—	—
	NbO_2F	TaO_2F	NbO_2Cl	TaO_2Cl	—	—	NbO_2I	TaO_2I
	Nb_3O_7F	Ta_3O_7F	Nb_3O_7Cl	—				
	$Nb_5O_{12}F$	—						
	$Nb_{17}O_{42}F$	—						
	$Nb_{31}O_{77}F$	—						
IV	NbF_4	—	$NbCl_4$	$TaCl_4$	$NbBr_4$	$TaBr_4$	NbI_4	TaI_4
	—	—	$NbOCl_2$	$TaOCl_2$	—	—	$NbOI_2$	—
III	NbF_3†	TaF_3†	$NbCl_3$	$TaCl_3$	$NbBr_3$	$TaBr_3$	NbI_3	TaI_3
Non-	—	—	Nb_3Cl_8	—	Nb_3Br_8	—	Nb_3I_8	—
integral	Nb_6F_{15}	—	—	Ta_6Cl_{15}	—	Ta_6Br_{15}	—	Ta_6I_{15}
	—	—	Nb_6Cl_{14}	—	—	Ta_6Br_{14}	—	Ta_6I_{14}
	—	—	—	—	—	—	Nb_6I_{11}	—

* $NbFCl_4$ and $TaFCl_4$ are also known.
† Existence uncertain.

TABLE 6. SELECTED THERMODYNAMIC PROPERTIES FOR NIOBIUM AND TANTALUM HALIDES AND OXYHALIDES*

Compound	M.p. (°C)	B.p. (°C)	ΔH_f	ΔH_{vap}	ΔH_{sub}	ΔH_{diss}	ΔH_{fus}	ΔS°_f	ΔS_{vap}	ΔS_{sub}	ΔS_{diss}
			kcal $mole^{-1}$ (at 298°)					cal deg^{-1} $mole^{-1}$ (at 298°)			
NbF_5	79·0	234·0	−433·5	14·3	22·6	—	2·92	95·1	—	53·2	—
TaF_5	97·0	229·0	−455·0	13·0	—	—	—	90·4	—	—	—
$NbCl_5$	203·4	247·4	−190·5	15·2	20·6	28·3	9·15	58·6	24·3	41·2	47·0
$TaCl_5$	215·9	232·9	−205·2	13·1	22·8	33·6	9·95	59·4	25·9	46·4	31·3
$NbBr_5$	254·0	365·0	−133·0	22·4	27·7	—	—	73·0	31·6	47·3	—
$TaBr_5$	256·0	344·0	−143·0	14·7	25·4	—	10·9	73·0	23·8	44·0	—
NbI_5	—	—	−64·6	—	—	—	—	82·0	—	—	—
TaI_5	496	543	−70·0	18·1	19·7	—	1·6	—	—	—	—
$NbCl_4$	—	—	−166·0	—	29·4	28·3	—	—	—	41·2	47·0
$TaCl_4$	—	—	−168·0	—	32·5	27·0	—	—	—	40·4	48·3
$NbCl_3$	—	—	−139·0	—	—	—	—	—	—	—	—
$TaCl_3$	—	—	−130·5	—	—	41·9	—	—	—	—	58·6
$NbOCl_3$	—	—	−210·5	—	26·2	—	—	31·1	—	44·3	—
$TaOCl_3$	—	—	−213·3	—	—	—	—	—	—	—	—
$NbOBr_3$	—	—	−178·6	—	—	—	—	—	—	—	—
$NbOCl_2$	—	—	−185·0	—	—	—	—	—	—	—	—
$TaOCl_2$	—	—	−188·0	—	—	—	—	—	—	—	—
Ta_6Cl_{15}	—	—	−113·4	—	—	—	—	—	—	—	—
Nb_6Cl_{14}	—	—	−113·5	—	—	—	—	—	—	—	—
Nb_3Cl_8	—	—	−128·6	—	—	—	—	—	—	—	—

* Certain values for the niobium compounds are taken from the recent initial assessment by Lavrentev and Gerassimov[a] whilst the remaining values and those for the tantalum compounds have been selected from the results listed by Canterford and Colton[b].

[a] V. I. Lavrentev and Ya. I. Gerassimov, *Atomic Energy Review*, Special Issue, No. 2 (1968) 7.

[b] J. H. Canterford and R. Colton, Chapter 4 in *Halides of the Second and Third Row Transition Metals*, Wiley and Sons, London (1968).

discussed later (page 573). The non-integral valence state halides, which are known to contain trimeric or hexameric metal clusters, have received much attention in recent years and undoubtedly further examples of these classes of compounds will be discovered. Metal–metal bonding is also encountered amongst the tetrahalides, as shown by the structural and magnetic data discussed later.

The fact that the pentahalides of niobium and tantalum are stable is in marked contrast with the situation existing with vanadium for which element only the pentafluoride is stable. The pentahalides are predominantly covalent, the volatility decreasing from the fluorides to the iodides. Niobium and tantalum are the only elements, apart from protactinium[75], which form pentaiodides. Lower halides are generally obtained by reduction or thermal decomposition of the appropriate pentahalide, the ease of reduction increasing from fluoride to iodide, tantalum compounds being more difficult to reduce than their niobium analogues. By varying reaction conditions it is possible to obtain a variety of products from a given MCl_5+M reaction and the use of "temperature gradient" reactions, extensively exploited by Schäfer and his colleagues (for reviews and discussion, see refs. 72–74), has been instrumental in elucidating the complex halide chemistry of these elements.

For convenience selected thermodynamic properties[73, 74] for the halides and oxyhalides are listed together in Table 6 and vapour pressure data are given in Tables 7 and 8, the information in the last table referring to the thermal decomposition of lower halides and complex halides.

TABLE 7. VAPOUR PRESSURE DATA FOR NIOBIUM AND TANTALUM HALIDES[a]

Compound	$\log p_{mm} = A - B/T$		Temperature range, °C
	A	B	
NbF_5	8·439	2824	70–230
NbF_5	8·372	2777	100–250
TaF_5	8·524	2834	80–230
$NbCl_5$	$\log p_{atm} = 19{\cdot}3779 - 3{\cdot}2204 \log T - 5353/T - 7649/T^2$		77–110
$NbCl_5$	8·37	2870	210–254
$TaCl_5$	$\log p_{atm} = 19{\cdot}422 - 3{\cdot}220 \log T - 5347/T - 7649/T^2$		77–110
$TaCl_5$	8·68	2970	220–240
$NbBr_5$	12·5	5782	205–252
$NbBr_5$	9·784	4347	252–356
$TaBr_5$	12·51	5546	180–255
$TaBr_5$	8·07	3204	255–344
TaI_5	8·21	4300	380–490
TaI_5	7·67	3950	500–540
$NbCl_4$*	12·30	6870	304–374
$NbOCl_3$†	13·35	6394	225–329

* In the presence of $NbCl_5$ to prevent dissociation.

† Corrected for the vapour pressure of $NbCl_5$ produced by thermal decomposition of $NbOCl_3$.

[a] Values selected from data compiled by Canterford and Colton, *Halides of the Second and Third Row Transition Elements*, Wiley and Sons, London (1968).

TABLE 8. EQUILIBRIUM DISSOCIATION PRESSURE OF THE PENTAHALIDE OR OXYHALIDE OVER THERMALLY UNSTABLE COMPOUNDS[a]

Compound	M.p. °C	Product	$\log p_{mm} = A - B/T$		Temperature range, °C
			A	*B*	
$NbOCl_3$	—	$NbCl_5$	17·30	9264	225–329
$NbOCl_3$*	—	$NbCl_5$	10·386	6270	337–397
NbF_4*	—	NbF_5	8·9	5500	400–500
$NbCl_4$	—	$NbCl_5$	13·16	6190	200–320
$TaCl_4$	—	$TaCl_5$	13·2	5700	
$TaBr_4$*	—	$TaBr_5$	9·83	6145	300–327
$TaCl_3$	—	$TaCl_5$	12·815	9173	377–427
$TaBr_3$*	—	$TaBr_5$	9·83	6821	335–415
$NaNbCl_6$	444	$NbCl_5$	7·65	4300	225–300
$NaTaCl_6$	484	$TaCl_5$	9·15	4030	
$KNbCl_6$	396	$NbCl_5$	8·22	6050	340–450
$KTaCl_6$	410	$TaCl_5$	8·36	4820	200–340
$NaNbOCl_4$*	430	$NbOCl_3$	8·077	5376	315–380
$KNbOCl_4$	428	$NbOCl_3$	12·10	6530	300–420
$RbNbOCl_4$	526	$NbOCl_3$	11·10	5830	340–450
$CsNbOCl_4$	455	$NbOCl_3$	7·65	3640	370–430
$KTaOCl_4$ (s)	625	$TaCl_5$	6·99	4912	525–600
$KTaOCl_4$ (l)		$TaCl_5$	4·17	2400	625–700
$RbTaOCl_4$ (s)	675	$TaCl_5$	7·18	5238	550–675
$RbTaOCl_4$ (l)		$TaCl_5$	3·75	2030	700–775
$CsTaOCl_4$ (s)	700	$TaCl_5$	8·69	7230	625–700
$CsTaOCl_4$ (l)		$TaCl_5$	4·77	3400	700–800
K_2NbOCl_5	500	$NbOCl_3$	6·99	4352	460–700
Rb_2NbOCl_5	570	$NbOCl_3$	6·66	4590	515–740
Cs_2NbOCl_5	640	$NbOCl_3$	7·05	5180	600–780

* P. in atmospheres.
(s) solid; (l) liquid.
[a] Data selected from Canterford and Colton, *Halides of the Second and Third Row Transition Elements*, Wiley and Sons, London (1968).

Polynuclear Halides

Hexameric cluster compounds of the types M_6X_{14} (M = Nb and Ta; X = Cl and Br) have been known for a considerable time[73, 74]. More recently cluster compounds of the type M_6X_{15} (X = F, Cl, Br and I) and the iodide Nb_6I_{11} have been characterized[77–80] and it is now suggested that the hydrated compounds originally formulated as $M_6X_{14}.7H_2O$ really contain *eight* molecules of water[81–83]. In addition, $Ta_6Br_{15}.8H_2O$ and hydrates of the types $M_6Cl_{14}.9H_2O$ and $M_6Cl_{15}.7H_2O$, have been reported[81]. The anhydrous compounds are conveniently prepared by reduction of higher halides with niobium, tantalum or aluminium (see refs. 72–74 and Table 9) and the hydrates form on crystallization of aqueous extracts

[77] H. Schäfer, H. G. von Schnering, K.-J. Niehues and H. G. Nieder-Varenholz, *J. Less-Common Metals*, **9** (1965) 95.
[78] D. Bauer and H. Schäfer, *J. Less-Common Metals*, **14** (1968) 476.
[79] D. Bauer and H. G. von Schnering, *Z. anorg. und allgem. Chem.* **361** (1968) 259.
[80] A. Simon, H. Schäfer and H. G. von Schnering, *Z. anorg. und allgem. Chem.* **355** (1967) 295.
[81] B. Spreckelmeyer, *Z. anorg. und allgem. Chem.* **358** (1968) 147.
[82] H. Schäfer and D. Bauer, *Z. anorg. und allgem. Chem.* **340** (1965) 62.
[83] P. B. Fleming, L. A. Mueller and R. E. McCarley, *Inorg. Chem.* **6** (1967) 1.

TABLE 9. PREPARATION OF ANHYDROUS, HEXAMERIC NIOBIUM AND TANTALUM HALIDES

Compound	Colour	Method	Reference
Nb_6F_{15}	Brown	NbF_5/Nb, temperature gradient, 400–900°	a
Ta_6Cl_{15}	Black	$TaCl_5$/Ta, temperature gradient, 470–630°	b
		$TaCl_5$/Al, temperature gradient, 200–400°	c
Ta_6Br_{15}	Black	$TaBr_5$/Ta, temperature gradient, 330–620°	d
Ta_6I_{15}	Black	$[Ta_6I_{12}]I_{18}$ heated at 100–200° in a vacuum	e
Nb_6Cl_{14}	Black	Nb_3Cl_8/Nb at 840°	f
Ta_6Br_{14}	Green	$TaBr_5$/Al, temperature gradient, 280–450°	c
Ta_6I_{14}	Black	TaI_5/Al, temperature gradient, 300–475°	c
		TaI_5/Ta, temperature gradient, 575–630°	g
Nb_6I_{11}	Black	Nb_3I_8/Nb, temperature gradient, 300–800°	h
		Niobium iodide/Nb at 1000°	i

[a] H. Schäfer, H. G. von Schnering, K. J. Niehues and H. G. Nieder-Vahrenholz, *J. Less-Common Metals*, **9** (1965) 95.
[b] H. Schäfer, H. Scholz and R. Gerken, *Z. anorg. und allgem. Chem.* **331** (1964) 154.
[c] P. J. Kuhn and R. E. McCarley, *Inorg. Chem.* **4** (1965) 1482.
[d] H. Schäfer, R. Gerken and H. Scholz, *Z. anorg. und allgem. Chem.* **335** (1965) 96.
[e] D. Bauer and H. Schäfer, *J. Less-Common Metals*, **14** (1968) 476.
[f] A. Simon, H. G. von Schnering, H. Wöhrle and H. Schäfer, *Z. anorg. und allgem. Chem.* **339** (1965) 155.
[g] D. Bauer, H. G. von Schnering and H. Schäfer, *J. Less-Common Metals*, **8** (1965) 388.
[h] H. Schäfer, H. G. von Schnering, A. Simon, D. Giegling, D. Bauer, R. Siepmann and B. Spreckelmeyer, *J. Less-Common Metals*, **10** (1966) 154.
[i] L. R. Bateman, J. F. Blount and L. F. Dahl, *J. Amer. Chem. Soc.* **88** (1966) 1082.

of initial reduction products. Some of the anhydrous compounds occur as equilibrium phases in the $TaBr_5$–Ta and TaI_5–Ta phase systems, but the only cluster compound observed in the NbI_5–Nb phase diagram is Nb_3I_8 which is discussed below.

The first evidence for the existence of the $[M_6X_{12}]^{2+}$ groups was reported by Vaughan *et al.*[84] who studied the scattering of X-rays by ethanolic solutions of certain of the complexes and deduced that an octahedron of metal atoms was present. Subsequent single-crystal studies have confirmed that the basic structural unit M_6X_{12} is present in Nb_6F_{15}, Ta_6Cl_{15}, Nb_6Cl_{14}, Ta_6I_{14} and $Ta_6Cl_{14}.7H_2O$. In Nb_6F_{15} [77] and Ta_6Cl_{15} [79] the metal atoms are arranged as a regular octahedron (Nb–Nb = 2·80 and Ta–Ta = 2·92 Å) surrounded by twelve bridging halogens arranged as a slightly deformed cuboctahedron. The M_6X_{12} groups are linked together by sharing the remaining halogens. Increasing distortion of the M_6 octahedron is found for Nb_6Cl_{14} and Ta_6I_{14}, respectively. This takes the form of a flattening of the octahedron (tetragonal distortion) as illustrated by the structure of the $[Ta_6I_{12}]$ group of Ta_6I_{14} [85] (Fig. 1). Metal–metal contacts within the octahedron are compared in Table 10. The distortion of the metal octahedron leads in turn to distortion of the coordination spheres of the two types of metal atom M(1) and M(2). Thus for Ta_6I_{14} the coordination around Ta(1) and Ta(2) is approximately octahedral and square pyramidal, respectively. In the M_6X_{14} class of compound the sixth octahedral coordination positions of the M(2) atoms are occupied by halogen atoms which each bridge to a neighbouring cluster, whilst for Nb_6F_{15} and Ta_6Cl_{15} additional inter-cluster bridging halogen atoms are bonded to each type of metal atom.

[84] P. A. Vaughan, J. H. Sturdivant and L. Pauling, *J. Am. Chem. Soc.* **72** (1950) 5477.
[85] D. Bauer, H. G. von Schnering and H. Schäfer, *J. Less-Common Metals*, **8** (1965) 388.

In contrast to the above flattening, the octahedron of metal atoms in $Ta_6Cl_{14}.7H_2O$ is reported to be elongated. Burbank[86] discusses the effect of the addition of extra ligands in excess of the composition M_6X_{12} pointing out that the additional ligands occupy vacant octahedral sites around the metal atoms and that the distortions of the tantalum octahedra in compounds of the type M_6X_{14} are more pronounced than the corresponding niobium distortions. Although solution spectral evidence supports the latter deduction it would be interesting to have further structural information in order to assess the effect of ligand size on the distortions in a given series of solid compounds such as, for example, M_6Cl_{14}, M_6Br_{14} and M_6I_{14} (M = Nb and Ta).

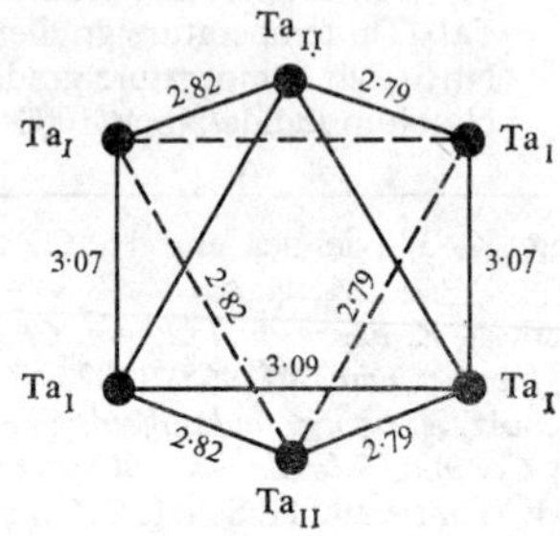

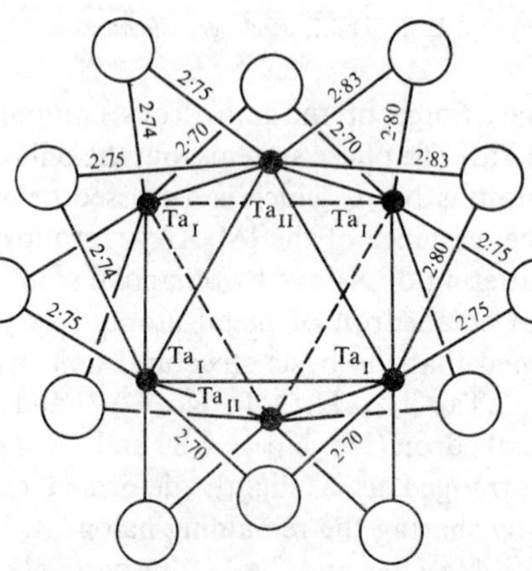

FIG. 1. The structure of the $[Ta_6I_{12}]$ group in Ta_6I_{14} [85].
(Reproduced by permission of the copyright holder.)

The hexameric unit $[Nb_6I_8]^{3+}$ in Nb_6I_{11} possesses a similar geometrical configuration to the molybdenum and tungsten hexamers with the remaining coordination site on each niobium atom occupied by an iodine atom which bridges to an adjacent cluster[80]. The niobium–niobium distances within the distorted Nb_6 group range from 2·72 to 2·96 Å. Absorption of hydrogen by Nb_6I_{11} above 300° leads to the formation of HNb_6I_{11} in which the hydrogen atom is located at the centre of the Nb_6I_{11} group[87]. Unlike Nb_6I_{11} this compound is diamagnetic.

[86] R. D. Burbank, *Inorg. Chem.* **5** (1966) 1491.
[87] A. Simon, *Z. anorg. und allgem. Chem.* **355** (1967) 311.

TABLE 10. CRYSTALLOGRAPHIC DATA FOR NIOBIUM AND TANTALUM HEXAMERIC CLUSTER COMPOUNDS

Compound	Symmetry	Space group	Lattice parameters (Å)			Metal–metal distances (Å)†		Reference
			a_0	b_0	c_0	M(1)–M(1)	M(1)–M(2)	
*Nb_6F_{15}	Cubic	O_h^9–$Im3m$	8·19	—	—	2·80	2·80	a
*Nb_6Cl_{14}	Orthorhombic	D_{2h}^{18}–$Bbam$	12·252	11·019	13·494	2·95–2·96	2·89	b
*Ta_6I_{14}	Orthorhombic	D_{2h}^{18}–$Bbam$	14·445	12·505	15·000	3·07–3·09	2·79–2·82	c
*$Ta_6Cl_{14}.7H_2O$	Trigonal	D_{3d}^1–$P\bar{3}1m$	9·36	—	8·80	—	—	d
*Nb_6I_{11}	Orthorhombic	D_{2h}^{10}–$Pccn$	11·299	15·309	13·558	—	—	e
Ta_6I_{15}	Cubic	O_h^9–$Im3m$	11·02	—	—	—	—	f
*Ta_6Cl_{15}	Cubic	O_h^{10}–$Ia3d$	20·28	—	—	2·92	2·92	g
Ta_6Br_{15}	Cubic	O_h^{10}–$Ia3d$	21·29	—	—	—	—	g
*$K_4Nb_6Cl_{18}$	Monoclinic	C_{2h}^3–$C2m$	9·97	16·51 β = 115°	9·47	2·91	2·91	h
*$(NMe_4)_2Nb_6Cl_{18}$	Trigonal	D_{3d}^3–$P\bar{3}ml$	10·85	—	8·615	—	—	i

* Structural data are given in references quoted.
† Distances are those between atoms labelled M(1) and M(2) in Fig. 1.
[a] H. Schäfer, H. G. Schnering, K. J. Niehues and H. G. Nieder-Varenholz, *J. Less-Common Metals*, **9** (1965) 95.
[b] A. Simon, H. G. Schnering, H. Wohrle and H. Schäfer, *Z. anorg. und allgem. Chem.* **339** (1965) 155.
[c] D. Bauer, H. G. von Schnering and H. Schäfer, *J. Less-Common Metals*, **8** (1965) 388.
[d] R. D. Burbank, *Inorg. Chem.* **5** (1966) 1491.
[e] A. Simon, H. Schäfer and H. G. von Schnering, *Z. anorg. und allgem. Chem.* **355** (1967) 295.
[f] D. Bauer and H. Schäfer, *J. Less-Common Metals*, **14** (1968) 476.
[g] D. Bauer and H. G. von Schnering, *Z. anorg. und allgem. Chem.* **361** (1968) 259.
[h] V. A. Simon, H. G. von Schnering and H. Schäfer, *Z. anorg. und allgem. Chem.* **361** (1968) 235.
[i] F. W. Kokinat and R. E. McCarley, Abstr. No. 118, 157th Meeting Amer. Chem. Soc., April 1969.

The nature of the bonding in the $[M_6X_{12}]^{2+}$ clusters is discussed by Cotton and Haas[88] using the molecular-orbital approach and the assumption that each metal atom is surrounded by a square-planar arrangement of halogens. However, in view of the subsequent structural studies discussed above it is apparent that this simple treatment is not entirely satisfactory. Spectral assignments based on this approach have recently been discussed[89].

Ta_6Cl_{14} and Nb_6Br_{14} dissolve in water and ethanol, but Nb_6F_{15} and Ta_6Cl_{15} are unaffected by dilute mineral acids. Nb_6F_{15}, the first metal-cluster fluoride known, decomposes to give niobium pentafluoride and niobium metal at 700° in a vacuum.

Cluster compounds containing the $[M_6X_{12}]^{2+}$ cation (X = Cl and Br) are oxidized in aqueous sulphuric acid by a variety of reagents such as iodine, chlorine, oxygen (see ref. 74), vanadium(V)[90] and cobalt(III)[91]. The cluster unit is not destroyed and the resulting species, $[M_6X_{12}]^{3+}$ or $[M_6X_{12}]^{4+}$, are reduced to $[M_6X_{12}]^{2+}$ species by chromium(II) or vanadium(II).

88 F. A. Cotton and T. E. Haas, *Inorg. Chem.* **3** (1964) 10.
89 W. van Bronswyk, *J. Chem. Soc.* (A) (1968) 692.
90 J. H. Espenson, *Inorg. Chem.* **7** (1968) 631.
91 J. H. Espenson and D. J. Boone, *Inorg. Chem.* **7** (1968) 636.

Polynuclear complexes of the types $(NEt_4)_n[(M_6Cl_{12}).Cl_6]$, where $n = 2$, 3 and 4, $(NEt_4)_2[(Nb_6Cl_{12}).X_4.(C_2H_5OH)_2]$, where X = Cl and Br, and $[Nb_6Cl_{12}[.X_2.nH_2O$, where X = Cl, Br, I and OH are known and the preparation of complexes with oxygen donor ligands, such as $(Nb_6Cl_{12}).Cl_2.L_4$, where, for example, L = dimethylsulphoxide and triphenylphosphine oxide has been reported[92]. $K_4[(Nb_6Cl_{12})Cl_6]$ is obtained[83] by heating Nb_3Cl_8 with potassium chloride or by niobium reduction of the pentachloride, mixed with potassium chloride. Structural information shows that the Nb_6 group retains the octahedral arrangement in this compound[93] (Nb–Nb distance = 2·91 Å) and for $(NMe_4)_2[(Nb_6Cl_{12})Cl_6]$[94] it is reported to be almost octahedral (mean Nb–Nb distance = 3·02 Å).

TABLE 11. CALCULATED AND OBSERVED INFRARED VIBRATIONS FOR CERTAIN HEXAMERIC NIOBIUM AND TANTALUM CLUSTER COMPOUNDS[a], *

Compound		M–M† ‡	$(M–X^i)_1$	$(M–X^i)_2$	X^i–M–X^i	M–X^a	X^i–M–X^a
$[Nb_6F_{12}]F_3$	obs.	—	515	448	410	754	200
	calc.	109	515	446	406	769	226
$[Nb_6Cl_{12}].Cl_2$	obs.	93	343	280	242	206	144†
	calc.	97	344	278	240	208	142
$[Nb_6Br_{12}]Br_2.8H_2O$	obs.	—	267	198	145†	—	108
	calc.	78	268	185	149	152	94
$[Ta_6Cl_{12}].Cl_3$	obs.	—	328	280	235	192	142†
	calc.	78	319	275	235	195	141
$[Ta_6Br_{12}].Br_2$	obs.	62	230	182	147†, 157	133	84, 107
	calc.	67	228	183	153	131	93
$[Ta_6I_{12}]I_2$	obs.	55, 57	181	147†	115	98	—
	calc.	59	196	141	111	100	72

* Observed bands for other cluster compounds are listed in ref. a.
† Bands in the 141–145 cm^{-1} region of the spectra of cluster compounds are assigned by R. A. Mackay and R. F. Schneider (*Inorg. Chem.* **7** (1968) 455) to metal–metal stretching vibrations. The remaining assignments given by these authors, who recorded spectra to 50 cm^{-1}, are in agreement with those tabulated.
‡ Force constants for the metal–metal bonds are less than 0·3 mdyne/Å.
X^i, bridging halogen; X^a, terminal halogen; M, niobium or tantalum.
[a] R. Mattes, *Z. anorg. und allgem. Chem.* **364** (1969) 297.

Anhydrous halides of the type M_6X_{14} and compounds containing the $[Nb_6X_{12}]^{4+}$ cation are diamagnetic whilst those containing the $[Nb_6X_{12}]^{3+}$ cation are paramagnetic. The e.s.r. sprectrum of $(NEt_4)_3[(Nb_6Cl_{12}).Cl_6]$ has been examined, the g value being 1·95. Infrared spectral results[95] for several niobium and tantalum hexameric cluster compounds are summarized in Table 11. At present there is a difference of opinion concerning the assignment of the metal–metal stretching vibration (141–145 cm^{-1} or 59–109 cm^{-1}) which, in either case, is observed at a much lower frequency than the metal–metal vibration in molybdenum hexameric cluster compounds (230 cm^{-1}). The force constants for the metal–metal bonds in the niobium and tantalum halides are less than 0·3 mdyne/Å[95].

[92] R. A. Field and D. L. Kepert, *J. Less-Common Metals*, **13** (1967) 378.
[93] A. Simon, H. G. von Schnering and H. Schäfer, *Z. anorg. und allgem. Chem.* **361** (1968) 235.
[94] F. W. Kokinat and R. E. McCarley, Abs. 157th Meeting Amer. Chem. Soc., April 1969, Abs. No. 179.
[95] R. Mattes, *Z. anorg. und allgem. Chem.* **364** (1969) 279.

Trimeric cluster compounds of the type M_3X_8 (X = Cl, Br and I) are known for niobium but not for tantalum. These are obtained by thermal decomposition of higher halides or by reduction of the pentahalide with niobium in a temperature gradient. β-forms of Nb_3Br_8 and Nb_3I_8 are crystallized by chemical transport reactions following preparation from the elements at temperatures in excess of 500°. Nb_3Cl_8 and Nb_3Br_8 are the limiting niobium-rich phases in the homogeneous phases extending over the ranges of composition $NbCl_{2\cdot67}$ to $NbCl_{3\cdot13}$ and $NbBr_{2\cdot67}$ to $NbBr_{3\cdot03}$, respectively.

Crystallographic data are listed in Table 12. The structures of both the α- and β-forms contain Nb_3-triangles, the metal–metal bond distances ranging between 2·81 and 3·00 Å. In β-Nb_3Br_8, for example[96], the niobium atoms are linked by bridging bromine atoms situated below the plane of the triangle and a fourth bridging bromine, situated above this plane, is shared by all three niobium atoms (Fig. 2). The remaining halogen atoms are involved in intercluster bridging.

Nb_3Cl_8, β-Nb_3Br_8 and β-Nb_3I_8 are paramagnetic, the magnetic moment for the first being 1·86 B.M. per Nb_3Cl_8 molecule. β-Nb_3Br_8 is more sensitive toward hydrolysis than the α-form and Nb_3I_8 is stable in air for several days.

TABLE 12. CRYSTALLOGRAPHIC DATA FOR THE NIOBIUM TRIMERIC CLUSTER COMPOUNDS

Compound	Colour	Symmetry	Space group	Lattice parameters (Å) a_0	Lattice parameters (Å) c_0	Nb–Nb distance (Å)	$\rho_{calc.}$ g cm^{-3}	Reference
*Nb_3Cl_8	green	Hexagonal	D^3_{3d}–$P\bar{3}ml$	6·744	12·268	2·81	3·87	a
α-Nb_3Br_8	black	Hexagonal	D^3_{3d}–$P\bar{3}ml$	7·227	12·930	—		b
*β-Nb_3Br_8	black	Hexagonal	D^5_{3d}–$R\bar{3}m$	7·080	38·975	2·88	5·404	c
*β-Nb_3I_8	black	Hexagonal	D^5_{3d}–$R\bar{3}m$	7·600	41·715	3·00	6·171	c

* Structural data are given in references cited.
[a] H. G. Schnering, H. Worhle and H. Schäfer, *Naturwissen*, **48** (1961) 159.
[b] S. S. Berdonosov and A. V. Lapitskii, *Russ. J. Inorg. Chem.* **10** (1965) 1525.
[c] A. Simon and H. G. Schnering, *J. Less-Common Metals*, **11** (1966) 31.

Trihalides

Although it is claimed that NbF_3 and TaF_3 have been prepared, more recent studies have shown that for niobium the products invariably contain oxygen. Cubic products of the general formula M[(O)F]$_3$ all possess the ReO_3-type structure, originally reported for NbF_3 and TaF_3, with small variations in lattice parameter. It is doubtful whether pure trifluorides have yet been prepared.

Trichlorides, tribromides and triiodides are obtained[72–74] by reduction of the pentahalide with hydrogen, aluminium or the appropriate metal or, in certain instances, by thermal decomposition of the tetrahalides. Reduction of $TaCl_5$ and $TaBr_5$ proceeds via the tetrahalides which then undergo disproportionation. Stoichiometric $NbCl_3$ and $NbBr_3$ are apparently only incidental compositions in the homogeneous phases $NbCl_{2\cdot67}$ to $NbCl_{3\cdot13}$

[96] A. Simon and H. G. von Schnering, *J. Less-Common Metals*, **11** (1966) 31.

and $NbBr_{2.67}$ to $NbBr_{3.03}$, respectively. $TaBr_3$, TaI_3 and NbI_3 are not observed as equilibrium phases in their respective pentahalide–metal phase diagrams. The last compound decomposes irreversibly at 513° into Nb_3I_8 and an iodine-rich liquid; it possesses hexagonal symmetry with $a_0 = 6{\cdot}61$ and $c_0 = 6{\cdot}82$ Å.

The thermally unstable trihalides are dark coloured, unreactive solids and their chemical properties have received little attention. $TaCl_3$, $NbBr_3$ and $TaBr_3$ are paramagnetic; the magnetic moment for both tribromides is 1·0 B.M.

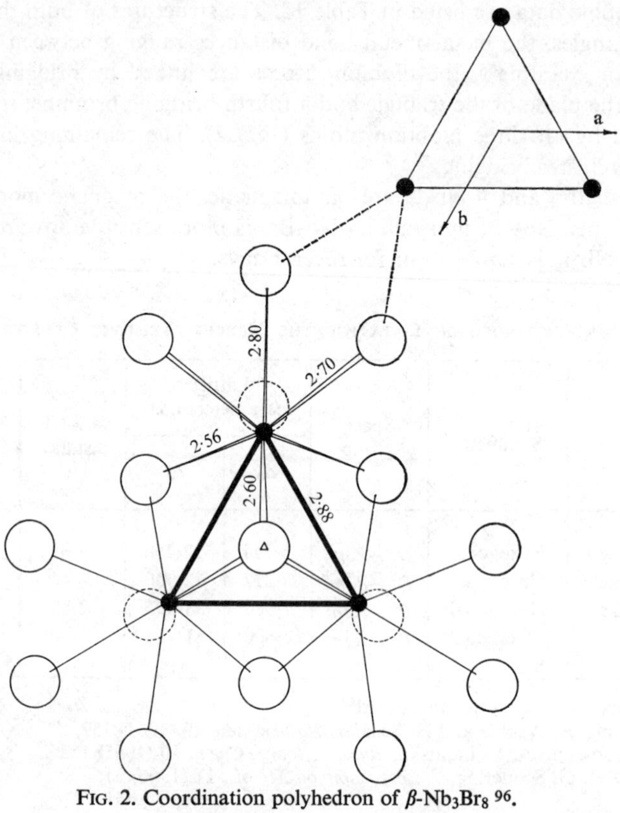

FIG. 2. Coordination polyhedron of β-Nb_3Br_8 [96].
●—Nb atoms; ○—Br atoms.
(Reproduced by permission of the copyright holder.)

Tetravalent Halides

Niobium tetrafluoride, a black, non-volatile, hygroscopic solid, is best prepared by reduction of the pentafluoride with niobium metal although other reagents such as silicon, phosphorus and boron may be used. The reduction of tantalum pentafluoride has not been achieved and recent evidence suggests that lower tantalum fluorides are thermodynamically unstable[97]. The structure of NbF_4 differs from that of the other *d*-transition element[74] and the actinide tetrafluorides[98]. The symmetry is tetragonal (Table 13) and each niobium atom is surrounded octahedrally by six fluorine atoms, four of which in a plane are shared by neighbouring niobiums resulting in the formation of a two-dimensional layer lattice[99].

[97] H. Schäfer and H. G. von Schnering, *Angew. Chem.* **76** (1964) 833.
[98] D. Brown, *Halides of the Lanthanides and Actinides*, Wiley and Sons, London (1968).
[99] F. P. Gortsema and R. Didchenko, *Inorg. Chem.* **4** (1965) 182.

Although metal–metal interactions are undoubtedly smaller than in the other tetrahalides the low, temperature-independent magnetic susceptibility of NbF_4, χ_m = 100–200 c.g.s. units, suggests that they are still important.

With the exception of niobium tetraiodide the remaining tetrahalides are conveniently prepared by reduction of the corresponding pentahalides, under controlled conditions at elevated temperatures, with either niobium or tantalum metal, aluminium or hydrogen. The ease of reduction of niobium pentachloride relative to its tantalum analogue is illustrated by the reactions with hydrogen and red phosphorus. Thus, hydrogen reduces niobium pentachloride at relatively low temperatures whereas tantalum pentachloride is only reduced at temperatures above 500° and, whilst red phosphorus reduces niobium pentachloride at 170° tantalum pentachloride is unaffected at temperatures up to 220°. Niobium tetraiodide is best prepared by the thermal decomposition of the pentaiodide in a sealed tube at 270°.

TABLE 13. CRYSTALLOGRAPHIC PROPERTIES OF THE TETRAVALENT HALIDES AND OXYHALIDES

Compound	Colour	Symmetry	Space group	Lattice parameters a_0 (Å)	b_0 (Å)	c_0 (Å)	β (°)	$\rho_{calc.}$ g cm^{-3}	Reference
*NbF_4	Black	Tetragonal	D_{4h}^{17}–I/mmm	4·082	—	8·161	—	4·13	a
$NbCl_4$	Violet-black	Monoclinic	C_{2h}^{3}–$C2/m$	12·32	6·82	8·21	134	3·23	b
$TaCl_4$	Black	Monoclinic	C_{2h}^{3}–$C2/m$	12·32	6·82	8·21	134	4·32	c
$NbCl_4$	Violet-black	Orthorhombic	—	8·12	8·88	6·84	—	3·23	d
$TaCl_4$	Black	Orthorhombic	—	8·16	8·92	6·80	—	4·32	e
$NbBr_4$	Black	Orthorhombic	—	8·60	9·31	7·19	—	4·77	d
$TaBr_4$	Brown-black	Orthorhombic	—	9·58	9·30	7·21	—	5·31	e
$NbBr_4$	Black	Orthorhombic	} *C2cm*, *A2₁am*, or *Amam*	7·179	12·22	12·85	—	4·86	f
$TaBr_4$	Brown-black	Orthorhombic	}	7·143	12·38	12·88	—	5·90	f
*α-NbI_4	Black	Orthorhombic	C_{2v}^{12}–$CmC2_1$	7·67	13·28	13·93	—	5·64	g
$NbOCl_2$	Black	Monoclinic	C_2^3–$C2$	12·79	3·93	6·70	105	—	b
$NbOI_2$	Black	Monoclinic	C_2^3–$C2$	15·08	3·92	7·48	105	—	b
*NbS_2Cl_2	Red-brown	Monoclinic	C_{2h}^{3}–$F2/m$	12·60	11·11	6·30	96·9	3·45	h

* Structural data given in the references quoted.
[a] F. P. Gortsema and R. Didchenko, *Inorg. Chem.* **4** (1965) 182.
[b] H. G. von Schnering and K. Wohrle, *Angew. Chem.* **2** (1963) 558.
[c] H. Schäfer, H. Scholz and R. Gerken, *Z. anorg. und allgem. Chem.* **331** (1964) 154.
[d] R. E. McCarley and B. A. Torp, *Inorg. Chem.* **2** (1963) 540.
[e] R. E. McCarley and J. C. Boatman, *Inorg. Chem.* **2** (1963) 547.
[f] S. S. Berdonosov, A. V. Lapitskii and D. G. Berdonosova, *Russ. J. Inorg. Chem.* **9** (1964) 1388.
[g] L. F. Dahl and D. L. Wampler, *Acta Cryst.* **15** (1962) 903.
[h] H. G. von Schnering and W. Beckmann, *Z. anorg. und allgem. Chem.* **347** (1966) 231.

Thermodynamic data are summarized in Table 6 and the available crystallographic properties are listed in Table 13. There appears to be some conflict of opinion concerning the symmetry of the tetrachlorides and tetrabromides although most cells are based on powder results and the single-crystal data (Table 13, ref. b) are probably most reliable. However, with the possible exception of the molybdenum and tungsten compounds, they are not isostructural with their *d*- and *f*-transition element analogues. Niobium and tantalum tetraiodides are polymorphic. The structure[100] of α-NbI_4 is based on a distorted hexagonal close-packing of iodine atoms with one-quarter of the available octahedral holes occupied by niobium atoms so as to give an infinite linear chain of NbI_6 octahedra sharing opposite edges. Successive niobium atoms are shifted in opposite directions from the centre of the octahedron to form pairs of metal atoms 3·31 Å apart. This structure is quite different from that of ThI_4 in which the thorium atoms are eight-coordinate[101]. It is briefly reported[102] that $NbCl_4$ has a fibrous structure with a metal–metal distance of 3·06 Å. Apart from NbF_4, mentioned above, the tetrahalides are all diamagnetic.

Thermal decomposition of the tetrahalides is complex[73, 74] and depending on the temperature the appropriate pentahalide is obtained together with either the trihalide, a lower halide such as Nb_6F_{15} or Nb_3I_8, or even the metal. The equilibrium vapour pressures of the resulting pentahalides at certain temperatures are listed in Table 8. The thermal instability of niobium and tantalum tetrahalides is paralleled by similar behaviour for many of the other *d*-transition element tetrahalides but contrasts with the stability of protactinium and most other actinide tetrahalides.

The tetrahalides are moisture-sensitive compounds which are hydrolysed in air, NbF_4 forming NbO_2F. Apart from NbF_4, for which no other chemical properties have been reported, they form complexes[73, 103] of the types $MX_4.2L$ (X = variously Cl, Br and I; L = methyl cyanide, pyridine, γ-picoline and 1:4 dioxane), $NbX_4.2L$ (X = Cl and Br; L = tetrahydrofuran and tetrahydropyran), NbX_4.2diarsine (X = Cl, Br and I) and $MX_4.L$ (L = 2,2′ bipyridyl and 1:10 phenanthroline), certain of which may be obtained by reduction of the pentahalide by the ligand (page 584). Electronic spectra have been recorded for several of the above-mentioned complexes and metal–halogen vibrations are observed in their infrared spectra in the ranges 340–240 cm^{-1} (M–Cl) and 265–215 cm^{-1} (M–Br)[103]. In contrast to the tetrahalides these complexes are paramagnetic indicating that the metal–metal interactions are destroyed on complex formation[73, 103]. The magnetic moments of the *o*-phenylene bis-(dimethylarsine) complexes (1·8–2·0 B.M.) are close to the spin-only value (1·73 B.M.) for one unpaired electron. These compounds are isostructural with other tetravalent diarsine complexes. In contrast to this behaviour diamagnetic complexes of the type $NbCl_4.L$ (L = triethylamine and NNN′N′-tetramethylenediamine) have recently been reported[104].

Tetravalent Oxyhalides

The diamagnetic oxydihalides $NbOCl_2$, $TaOCl_2$ and $NbOI_2$ are obtained by heating mixtures of the metal and the pentoxide with either the pentachloride (350–400°) or iodine (500°). $NbOCl_2$ is alternatively prepared by reducing $NbOCl_3$ with hydrogen or stannous

100 L. F. Dahl and D. L. Wampler, *Acta Cryst.* **15** (1962) 903.
101 A. Zalkin, J. D. Forrester and D. H. Templeton, *Inorg. Chem.* **3** (1964) 639.
102 H. G. von Schnering and K. Wohrle, *Angew. Chem.* **2** (1963) 558.
103 G. W. A. Fowles, D. J. Tidmarsh and R. A. Walton, *Inorg. Chem.* **8** (1969) 631.
104 T. M. Brown and G. S. Newton, *Inorg. Chem.* **5** (1966) 1117.

chloride. They are non-volatile, air-stable solids which are stable towards dilute mineral acids. At high temperatures $NbOCl_2$ is reported[105] to decompose to give $NbOCl_3$ and $NbOCl$ (!) whereas $NbOI_2$ decomposes to yield a residue containing NbO_2 and NbO. These decompositions contrast with those of the actinide(IV) oxydihalides which disproportionate to form the dioxide and the volatile tetrahalide[98]. Crystallographic properties are listed in Table 13 and the heats of formation are given in Table 6. $MoOCl_2$ is isostructural with $NbOCl_2$ and $TaOCl_2$ but the actinide oxydichlorides[98] possess a different structure. Metal–metal distances for $NbOCl_2$, $NbOBr_2$ and $NbOI_2$ are 3·14, 3·12 and 3·16 Å, respectively[102, 106].

Tetravalent Thio- and Selenohalides

Tetravalent sulphide dihalides and selenide dihalides of general formula NbS_2X_2 and $NbSe_2X_2$ (X = Cl, Br and I) are prepared by combination of the component elements in a temperature gradient or in the case of NbS_2Cl_2 by the reaction between metallic niobium and S_2Cl_2 [107]. Thermal decomposition of the iodides results in the formation of niobium disulphide or diselenide and free iodine, but the remaining compounds decompose according to the equation

$$3NbS_2Cl_2 \rightarrow NbS_2 + S_{2(g)} + 2NbSCl_{3(g)}$$

The gaseous species $NbSCl_3$, $NbSBr_3$, $NbSeCl_3$ and $NbSeBr_3$ were identified by mass spectroscopy.

The tetravalent compounds are all diamagnetic and the presence of Nb–Nb bonding has been confirmed by structural studies on NbS_2Cl_2, which possesses monoclinic symmetry (Table 13), space group C_{2h}^3–$Fm3m$[106]. Discrete S_2 groups (S–S = 2·03 Å) are bonded to each niobium atom (Nb–Nb = 2·90 Å) and the $Nb_2(S_2)_2$ groups are linked together by chlorine atoms. The Nb–S bond distances are Nb–S_1, 2·49 Å and Nb–S_2, 2·51 Å, the Nb–Cl distances being 2·60 Å (×2) and 2·62 Å (×2).

Pentavalent Halides

All the pentahalides are most conveniently obtained by direct combination of the elements at elevated temperatures[72, 73]. Care is necessary during the preparation of NbI_5 because of the ready thermal decomposition of this compound. Alternative methods are discussed in detail elsewhere[72–74]: useful ones include reaction of the metals with chlorine trifluoride (pentafluorides), the reaction between niobium hydroxide and liquid thionyl chloride at room temperature[108] ($NbCl_5$) and the metathesis of the pentachlorides using boron tribromide[109] (pentabromides). It is also likely that reagents such as silicon tetraiodide will convert the pentachlorides to the pentaiodides as described for the preparation of PaI_5 [75]. Apart from NbI_5 the pentahalides are purified by vacuum sublimation; vapour pressure data are listed in Table 7.

Thermodynamic data are listed in Table 6 and crystallographic properties, together with calculated densities, are given in Table 14. Equations showing the temperature dependence of liquid densities are given in Table 15. The viscosities of the pentafluorides at their melting points are 91·41 and 70·31 cp for NbF_5 and TaF_5, respectively, indicating that the liquids are

105 I. S. Morozov and N. P. Lipatova, *Russ. J. Inorg. Chem.* **11** (1966) 550.
106 H. G. Schnering and W. Beckmann, *Z. anorg. und allgem. Chem.* **347** (1966) 231.
107 H. Schäfer and W. Beckmann, *Z. anorg. und allgem. Chem.* **347** (1966) 225.
108 D. Brown, *Inorg. Syn.* **8** (1967) 88.
109 P. M. Druce, M. F. Lappert and P. N. K. Riley, *Chem. Commun.* (1967) 486.

TABLE 14. CRYSTALLOGRAPHIC DATA FOR THE PENTAVALENT HALIDES AND OXYHALIDES

Compound	Colour	Symmetry	Space group	Lattice parameters					Reference
				a_0 (Å)	b_0 (Å)	c_0 (Å)	β (°)	$\rho_{calc.}$ g cm^{-3}	
*NbF_5	White	Monoclinic	C^3_{2h}–$C2/m$	9·62	14·43	5·12	96·1	3·54	a
*TaF_5	White	Monoclinic	C^3_{2h}–$C2/m$	9·64	14·45	5·12	96·3	5·19	a
*$NbCl_5$	Yellow	Monoclinic	C^3_{2h}–$C2/m$	18·30	17·98	5·888	90·6	2·78	b
$TaCl_5$	White	Monoclinic	C^3_{2h}–$C2/m$	18·30	17·98	5·888	90·6	3·76	b
$NbBr_5$	Red	Monoclinic	C^3_{2h}–$C2/m$	19·20	18·60	6·00	90·0	—	b
$NbBr_5$	Red	Orthorhombic	D^9_{2h}–$Pbam$	6·125	12·92	18·60	—	4·44	c
$TaBr_5$	Orange	Orthorhombic	D^9_{2h}–$Pbam$	6·125	12·92	18·60	—	5·24	c
$NbBr_5$	Red	Orthorhombic	D^9_{2h}–$Pbam$	6·127	12·98	18·55	—	4·44	d
$TaBr_5$	Orange	Orthorhombic	D^9_{2h}–$Pbam$	6·155	13·29	18·66	—	5·24	d
*NbI_5	Brass-coloured	Monoclinic	C^5_{2h}–$P2_1/c$	10·58	6·58	13·88	109·1	5·29	e
TaI_5	Black	Orthorhombic	—	6·65	13·95	20·10	—	5·809	f; g
*NbO_2F	White	Cubic	O^1_h–$Pm3m$	3·902	—	—	—	—	h
*TaO_2F	White	Cubic	O^1_h–$Pm3m$	3·898	—	—	—	—	h
*Nb_3O_7F	White	Orthorhombic	D^{19}_{2h}–$Cmmm$	20·67	3·833	3·927	—	4·37	i; o
Ta_3O_7F	White	Orthorhombic	—	16·690	3·935	8·915	—	—	j
Ta_3O_7F	White	Orthorhombic	—	6·478	10·496	3·907	—	—	j
$Nb_5O_{12}F$	White	Orthorhombic	—	6·15	18·29	3·92	—	5·09	i
*$Nb_{17}O_{42}F$	White	Monoclinic	C^1_2–$P2$	21·09	3·827	23·02	116·2	—	k
*$Nb_{31}O_{77}F$	White	Monoclinic	C^3_2–$C2$	37·54	3·832	21·18	91·92	—	k
*$NbOCl_3$	White	Tetragonal	D^{14}_{4h}–$P4_2/mnm$	10·87	—	3·96	—	3·27	l
*Nb_3O_7Cl	Blue-black	Orthorhombic	D^{16}_{2h}–$Pmnb$	10·88	15·53	3·84	—	—	m
*$TaCl_4F$	Yellow	Tetragonal	S^2_4–$I\bar{4}$	12·71	—	7·84	—	3·57	n

* Structural data are given in the references listed.
[a] A. J. Edwards, *J. Chem. Soc.* (1964) 3714.
[b] A Zalkin and D. E. Sands, *Acta Cryst.* **11** (1958) 615.
[c] R. F. Rolsten, *J. Phys. Chem.* **62** (1958) 126.
[d] S. S. Berdonosov, A. V. Lapitskii, D. G. Berdonosova and L. G. Vlasov, *Russ. J. Inorg. Chem.* **8** (1963) 1315.
[e] W. Littke and G. Brauer, *Z. anorg. und allgem. Chem.* **325** (1963) 122.
[f] R. F. Rolsten, *J. Am. Chem. Soc.* **80** (1958) 2952.
[g] J. F. Easey, Ph.D. Thesis, Leicester University (1968).
[h] L. K. Frevel and H. W. Rinn, *Acta Cryst.* **9** (1956) 626.
[i] S. Andersson and A. Astrom, *Acta Chem. Scand.* **18** (1964) 2233.
[j] L. Janhberg and S. Andersson, *Acta Chem. Scand.* **21** (1967) 615.
[k] A. Astrom, *Acta Chem. Scand.* **20** (1966) 969.
[l] D. E. Sands, A. Zalkin and R. E. Elson, *Acta Cryst.* **12** (1959) 21.
[m] H. G. von Schnering and W. Merten, *Naturwiss.* **51** (1964) 552
[n] H. Preiss, *Z. anorg. und allgem. Chem.* **346** (1966) 272.
[o] S. Andersson, *Acta Chem. Scand.* **18** (1964) 2339.

TABLE 15. VARIATION OF PENTAHALIDE DENSITIES WITH TEMPERATURE

Compound	ρ (g cm^{-3})	Temperature range, °C	Reference
NbF_5	$2{\cdot}6955-2{\cdot}490\times10^{-3}\,(t-80)$	87·1–125·6	a
TaF_5	$3{\cdot}8804-4{\cdot}035\times10^{-3}t\,(t-95)$	109·2–130·4	a
$NbCl_5$	$2{\cdot}0737-3{\cdot}115\times10^{-3}\Delta t+3{\cdot}58\times10^{-3}\Delta t^2$	204·2–325	b
$TaCl_5$	$2{\cdot}6840-4{\cdot}100\times10^{-3}\Delta t$	216·2–305	c
$NbBr_5$	$3{\cdot}324-2{\cdot}92\times10^{-3}\Delta t+0{\cdot}81\times10^{-6}\Delta t^2$	255–410	d
$TaBr_5$	$3{\cdot}718-3{\cdot}70\times10^{-3}\Delta t+2{\cdot}48\times10^{-6}\Delta t^2$	268–413	d

[a] F. Fairbrother, K. H. Grundy and A. Thompson, *J. Chem. Soc.* (1965) 761.
[b] L. A. Nisel'son and A. I. Pustil'nik, *Izv. Akad. Nauk SSSR, Otd. Tekhn. Nauk, Meti Gorn* (1963) 110.
[c] L. A. Nisel'son, A. I. Pustil'nik and T. D. Sokolova, *Russ. J. Inorg. Chem.* **9** (1964) 574.
[d] L. A. Nisel'son and T. D. Sokolova, *Russ. J. Inorg. Chem.* **9** (1964) 1117.

highly associated, a fact supported by the high values of the Trouton constant, which are 25·4 and 25·9, respectively[110]. Specific conductivities at the melting points are $1{\cdot}63\times10^{-5}$ (NbF_5) and $1{\cdot}56\times10^{-5}$ (TaF_5) mhos; it is suggested that slight self-ionization (<1%), $2MF_5 \rightleftharpoons MF_4^- + MF_6^+$, is responsible for the conductivity of the melt. The specific conductivities of molten niobium and tantalum pentachlorides are appreciably lower, being $2{\cdot}2\times10^{-7}$ and $3{\cdot}0\times10^{-7}$ mhos, respectively, whilst their viscosities at their melting points are 0·921 and 1·003 cp, respectively.

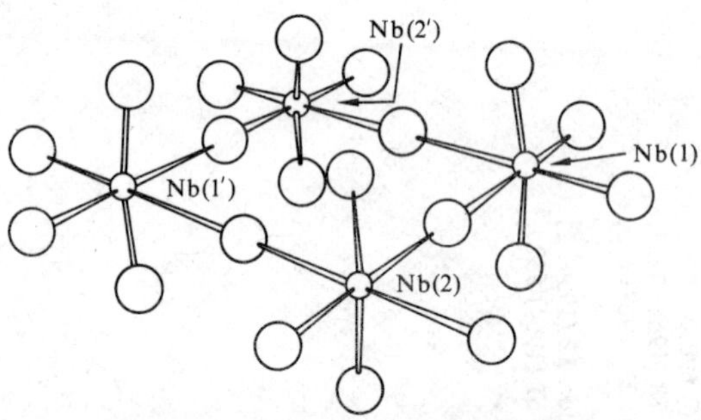

FIG. 3. The tetrameric unit of NbF_5 [111].
Average Nb–F = 1·77 and 2·06 Å to terminal and bridging atoms, respectively.
(Reproduced by permission of the copyright holder.)

Detailed structural information is available for the pentafluorides, the pentachlorides and the tetrachlorofluorides, MCl_4F, the last being obtained by the action of arsenic trifluoride on the pentachloride–phosphorus pentachloride adducts in arsenic trichloride.

Niobium and tantalum pentafluorides, like MoF_5, are tetramic in the solid state[111], the metal atoms at the corners of an almost perfect square being linked by four coplanar bridging atoms (Fig. 3). Average metal–fluorine bond lengths are 1·77 Å and 2·06 Å to terminal and bridging fluorine atoms, respectively. The pentafluorides of ruthenium, osmium, rhodium, iridium and platinum crystallize with a distorted MoF_5-type structure whereas those of uranium and protactinium possess entirely different structures. Similar tetrameric units are reported[112, 113] for niobium and tantalum tetrachlorofluorides in which the

110 F. Fairbrother, K. H. Grundy and A. Thompson, *J. Chem. Soc.* (1965) 761.
111 A. J. Edwards, *J. Chem. Soc.* (1964) 3714.
112 H. Preiss, *Z. anorg. und allgem. Chem.* **346** (1966) 272.
113 H. Preiss, *Z. anorg. und allgem. Chem.* **362** (1968) 13.

fluorine atoms are again bridging and coplanar with the metal atoms. The metal–fluorine distances vary between 2·10 and 2·11 Å with metal to chlorine bond lengths ranging from 2·26 to 2·31 Å. The fluorine bridges are symmetrical and almost linear in both the tetrafluorides and tetrachlorofluorides.

Niobium and tantalum pentachloride exist as dimeric units in the solid state. Each metal atom is surrounded by a distorted octahedron of chlorine atoms, two of which are bridging (Fig. 4(a)). Niobium–chlorine bond lengths vary from 2·25 to 2·30 Å for terminal atoms to

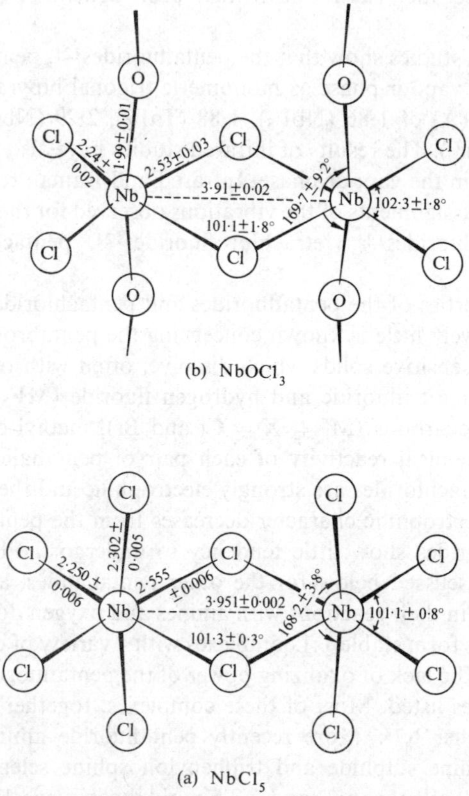

FIG. 4(a). The configuration of the Nb_2Cl_{10} dimeric molecule[134].
(b). The molecular configuration of $NbOCl_3$ [134].
(Reproduced by permission of the copyright holder.)

2·55 Å for bridging atoms[114]. In these compounds and the isostructural molybdenum and tungsten pentachlorides the chlorine atoms are in a hexagonal close-packing arrangement, whereas in $ReCl_5$ [115], which is also a dimer, they are in a double hexagonal close-packing array with the rhenium atoms occupying $\frac{1}{5}$ of the octahedral holes. Dimeric UCl_5 [116] and β-$PaBr_5$[117] provide yet another distinct way of forming M_2X_{10} dimers in the solid state since the halogen atoms in these compounds are in a cubic close-packing arrangement.

114 A. Zalkin and D. E. Sands, *Acta Cryst.* **11** (1958) 615.
115 K. Mucker, G. S. Smith and Q. Johnson, *Acta Cryst.* **B24** (1968) 874.
116 G. S. Smith, Q. Johnson and R. E. Elson, *Acta Cryst.* **22** (1967) 300.
117 D. Brown, T. J. Petcher and A. J. Smith, *Acta Cryst.* **B25** (1969) 178.

In contrast to these dimeric structures, protactinium pentachloride is unique in possessing a seven-coordinate, infinite chain structure with the metal atom at the centre of a pentagonal bipyramid[118].

The situation with respect to the pentabromides is to say the least confusing and, as shown in Table 14, X-ray powder data have been interpreted on the basis of different unit cells. During a recent attempt to resolve this problem by single-crystal studies all crystals obtained were found to be disordered[119]. Single-crystal studies have confirmed the unit cell originally reported for TaI_5, but it has been concluded that such crystals are twinned[120].

Electron diffraction studies show that the pentafluorides[121], pentachlorides[74] and pentabromides[74] exist in the vapour phase as monomeric trigonal bipyramids with mean metal–halogen bond lengths (Å) of 1·86 (NbF_5), 1·88 (TaF_5), 2·29 ($NbCl_5$), 2·30 ($TaCl_5$), 2·45 ($NbBr_5$) and 2·46 ($TaBr_5$). The results of infrared studies have also indicated that NbF_5 is a pentagonal bipyramid in the vapour phase. Infrared and Raman results have recently been reported together with assignments of the vibrations observed for the liquid pentafluorides[122] and the solid pentafluorides[123], tetrachlorofluorides[123], pentachlorides[124] and pentabromides[124].

The chemical properties of the pentafluorides and pentachlorides have been extensively investigated, but relatively little is known concerning the pentabromides and pentaiodides. They are all moisture-sensitive solids which dissolve, often with reaction, in non-aqueous solvents such as bromine trifluoride and hydrogen fluoride (MF_5), alcohols (MX_5; X = F → I inclusive), hydrocarbons (MX_5; X = Cl and Br), methyl cyanide (MX_5; X = Cl, Br and I), etc. The chemical reactivity of each pair of pentahalides is very similar. The pentafluorides and pentachlorides are strongly electrophilic and they catalyse Friedel–Craft type reactions. This electrophilic character decreases from the pentafluorides to the pentaiodides. The pentafluorides show little tendency to undergo the halogen elimination and reduction reactions discussed below for the other pentahalides, a characteristic which is particularly noticeable in their reactions with amines and oxygen donor ligands.

All the pentahalides form stable 1:1 complexes with a variety of donor ligands. Examples are listed in Table 16. The lack of oxidizing power of the pentafluorides is apparent from the nature of the complexes listed. Most of these complexes, together with several others, are discussed by Fairbrother[72, 73]. More recently pentafluoride–amine complexes[125], pentahalide–triphenylphosphine sulphide and triphenylphosphine selenide complexes[126], some pentahalide–thioxan[127], selenoxan[127] and 2,2′bipyridyl complexes[128] and the pentachloride complexes[129] $NbCl_5.CNCl$ and $TaCl_5.CNCl$ have been reported.

118 R. P. Dodge, G. S. Smith, Q. Johnson and R. E. Elson, *Acta Cryst.* **22** (1967) 85.
119 J. F. Easey, Private communication (1968).
120 J. F. Easey, Ph.D. Thesis, Leicester University (1968).
121 G. V. Romanov, V. P. Spinanov, *Izv. Sib. Otdel Nauk. SSSR, Ser. Khim. Nauk.* 126 (1968).
122 H. Seliz, A. Reis and E. L. Gasner, *J. Inorg. Nucl. Chem.* **30** (1968) 2087.
123 H. Preiss and P. Reich, *Z. anorg. und allgem. Chem.* **362** (1968) 19.
124 I. R. Beattie, T. R. Gilson and G. A. Ozin, *J. Chem. Soc.* (A) (1968) 2765.
125 Yu. A. Buslaev, M. A. Glushkova, M. M. Ershova and V. A. Bachhareva, *Russ. J. Inorg. Chem.* **13** (1968) 31.
126 D. Brown, J. Hill and C. E. F. Rickard, *J. Less-Common Metals* **20** (1970) 57.
127 K. L. Baker and G. W. A. Fowles, *J. Chem. Soc.* (A) (1968) 801.
128 G. W. A. Fowles, D. J. Tidmarsh and R. A. Walton, *J. Chem. Soc.* (A) (1969) 1546.
129 J. MacCordick, *Compt. Rend.* **266** (1968) 1296.

TABLE 16. EXAMPLES OF STOICHIOMETRIC 1:1 COMPLEXES FORMED BY NIOBIUM AND TANTALUM PENTAHALIDES

Ligand	X for complexes of the type $MX_5.L$ (M = Nb and Ta)
$(CH_3)_2O$	F, Cl and Br (F*)
$(C_2H_5)_2O$	F, Cl and Br
$(CH_3)_2S$	F, Cl and Br
$(C_2H_5)_2S$	F, Cl and Br
R_3PO (R = Ph and $N(CH_3)_2$)	Cl and Br
$(C_6H_5)_3PS$	Cl and Br
$(C_6H_5)_3PSe$	Cl and Br
C_4H_8S	F*, Cl* and Br*
C_4H_8OS	Cl
C_4H_8OSe	Cl
RCN (R = CH_3, C_2H_5 and n-C_3H_7)	Cl, Br and I
$NH(C_2H_5)_2$	F
$N(C_2H_5)_3$	F, Cl and Br
Diarsine	Cl and Br
BrF_3	F
XeF_2	F*, †
SeF_4	F
PCl_5	Cl
MCl_4 (M = S, Se and Te)	Cl
$POCl_3$	Cl
$(C_6H_5)_3X$ (X = N, P, As, Sb and Bi)	Cl‡ (M=Nb)

* Under certain conditions 1:2 complexes are also formed.
† Also 2:1 complexes known.
‡ Complexes of the type $2NbCl_5.(C_6H_5)X$ also known.

The new tantalum pentafluoride–xenon difluoride adducts $TaF_5.2XeF_2$ and $2TaF_5.XeF_2$ have been prepared by reactions in anhydrous hydrofluoric acid[130] and, together with the analogous niobium(V) complexes, by direct combination of the component halides[131]. Thermal decomposition of the 1:1 adducts results ultimately in the formation of the pentafluorides; xenon and fluorine are liberated as the decomposition proceeds with the intermediate formation of the 2:1 adducts. X-ray powder results suggests that $XeF_2.2TaF_5$ possesses a structure similar to that recently reported for $XeF_2.2SbF_5$ [132]. The stability of the Group VA pentafluoride–xenon difluoride complexes ($TaF_5 > NbF_5 > VF_5$) apparently depends on the Lewis acid strength of the pentafluoride.

A reinvestigation of the interaction of niobium pentafluoride and selenium tetrafluoride has shown[133] the major product to be $SeF_4.2NbF_5$. The complexes $SeF_4.NbF_5$ and $SeOF_2.NbF_5$ were identified as minor products of this reaction. The former, which is isostructural with $SeF_4.TaF_5$, possesses rhombohedral symmetry with $a_0 = 8{\cdot}99$ Å and $\alpha = 93°$, whilst the latter is orthorhombic with $a_0 = 8{\cdot}02$, $b_0 = 7{\cdot}13$ and $c_0 = 11{\cdot}09$ Å. $SeF_4.2NbF_5$ possesses monoclinic symmetry with $a_0 = 7{\cdot}60$, $b_0 = 17{\cdot}38$, $c_0 = 8{\cdot}80$ Å and $\beta = 103{\cdot}4°$, space group C^5_{2h}–$P2_1/c$. Structural studies have shown[133] that the molecular

130 W. J. Lawerence and G. D. Sturgeon, Abs. 157th Meeting Amer. Chem. Soc., April 1969, Abs. No. 137.
131 J. H. Holloway and J. G. Knowles, *J. Chem. Soc.* (A) (1969) 756.
132 V. M. McRae, R. D. Peacock and D. Russell, *Chem. Commun.* (1969) 62.
133 A. J. Edwards and G. R. Jones, *Chem. Commun.* (1968) 346.

geometry is consistent with the formulation (SeF_3^+) ($Nb_2F_{11}^-$), with substantial interaction between the ions through fluorine bridging as for $BrF_3.SbF_5$. The arrangement of the fluorine atoms around each niobium atom is approximately octahedral (Fig. 5).

The pentachloride and pentabromide–diethylether complexes are thermally unstable, yielding the corresponding oxytrihalides when heated in a vacuum. However, the analogous pentafluoride complexes can be sublimed unchanged, presumably as a consequence of the greater electronegativity of fluorine. The thioether complexes are appreciably more stable than the simple ether complexes, possibly due to the greater polarizability of the sulphur atom as compared with the oxygen atom. It is interesting to note in this connection that coordination via the sulphur atom is believed to take place in the thioxan complexes[127]. In the presence of an excess of ligands such as, for example, triphenylphosphine oxide, dimethylsulphoxide and diphenylsulphoxide oxygen abstraction occurs and oxyhalogeno-complexes of the types $MOCl_3.2Ph_3PO$, $Ta_2O_5Br.3(CH_3)_3SO$ and $TaO_2Br.2Ph_2SO$ are formed.

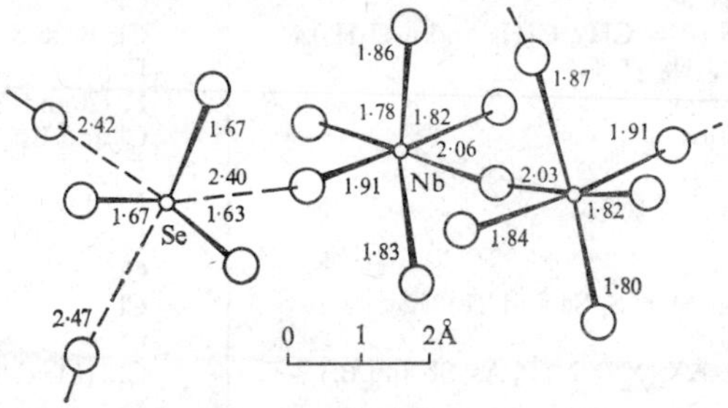

FIG. 5. The atomic arrangement in $SeF_4.2NbF_5$ projected down [001][138].
O—fluorine atoms.
(Reproduced by permission of the copyright holder.)

With mono-, di- and trialkylamines the pentafluorides form simple 1:1 adducts whereas with, for example, methylamine, dimethylamine and diethylamine the pentachlorides and pentabromides undergo replacement reactions to yield compounds of the types $MCl_2.(NHCH_3)_3$, $MCl_3.(NHR)_2.NH_2R$ and $MCl_3(NR_2).NHR_2$. With trimethylamine, however, 1:2 complexes are formed, viz. $NbCl_5.2N(CH_3)_3$, $TaCl_5.2N(CH_3)_3$ and $NbBr_5.2N(CH_3)_3$. Examples of other replacement reactions (listed in Table 17) are discussed later. There is also a trend in the reactions with pyridine. The pentafluorides form stable 1:2 complexes, whereas, depending on the reaction conditions, the other pentahalides either form 1:1 complexes ($NbCl_5$, $TaCl_5$ and $TaBr_5$) or undergo reduction to give tetrahalide complexes of the type $MX_4.2C_6H_5N$ (M = Nb and Ta; X = Cl, Br and I). Similar reduction products are obtained from reactions with γ-picoline and tetravalent 1:1 complexes result from reactions with 1,10-phenanthroline and, under certain conditions, with 2,2′-bipyridyl.

Pentavalent Oxyhalides

Of the oxytrihalides (Table 5) $NbOCl_3$ is undoubtedly the most extensively studied. This is due to the fact that it is formed during many methods directed at the preparation of niobium pentachloride and owing to their similar volatilities (Table 7) they are extremely difficult to separate. Direct methods of preparation include[73,74] the reaction between niobium pentachloride and oxygen at 300–500° and the thermal decomposition of the etherate $NbCl_5.O(C_2H_5)_2$. Analogous reactions have been used for the preparation of $NbOBr_3$,

TABLE 17. EXAMPLES OF REPLACEMENT REACTIONS INVOLVING NIOBIUM AND TANTALUM PENTAHALIDES

Reactants	Product
$NbCl_5$, NaC_5H_5	$(C_5H_5)_2NbCl_3$ and $(C_5H_5)_3NbCl_2$
$TaCl_5$, $NaC_5H_5+NaBH_4$ in tetrahydrofuran	$(C_5H_5)_2TaH_3$
MBr_5, NaC_5H_5 in tetrahydrofuran	$(C_5H_5)_2MBr_3$
MCl_5, NaC_5H_5 excess, in benzene	$M(C_5H_5)_4$
MCl_5, $Zn(CH_3)_2$	$(CH_3)_3MCl_2$
$NbCl_5$, $LiAlH_4$	$Nb(AlH_4)_x$, $x = 2{\cdot}5$–$3{\cdot}5$
MCl_5, $LiNR_2$ ($R = CH_3$ and C_2H_5)	$M(NR_2)_5$ or $NbN(C_2H_5)_4$
$NbCl_5$, $LiNR_2$ ($R = n$-C_3H_7, n-C_4H_9)	$Nb(NR_2)_4$
MCl_5, Acac, ROH ($R = CH_3$ or C_2H_5)	MCl_2.Acac$(OR)_2$
MCl_5, ROH ($R = C_6H_5$ and $C_{10}H_7$)	$M(OR)_5$, $M(OC_6H_5)_4Cl$, or $M(OC_6H_5)_3Cl_2$ depending on the conditions
MCl_5, $ROH+NH_3$ gas ($R = CH_3$, C_2H_5, n-C_3H_7 etc.)	$M(OR)_5$
MCl_5, salicylic acid	$M_2(OC_6H_4COO)_5$ or $M_2Cl_4.(OC_6H_4COO)_3$ depending on the conditions
MCl_5, NH_4Cl	$MNCl_2$
MCl_5, NH_3	$M(NH_2)_2Cl_3.7NH_3$

$TaOBr_3$ and $TaOCl_3$, but $NbOI_3$ is best prepared by the reaction of the pentoxide with NbI_5, or with a mixture of niobium and iodine in a temperature gradient. Although the existence of $TaOF_3$ has been reported, the oxytrifluorides have barely been studied.

The structure of solid $NbOCl_3$ consists of chains of dimeric Nb_2Cl_6 units linked by bridging oxygen atoms, the coordination around each niobium being approximately octahedral[134]. Bond distances and angles are shown in Fig. 4(b) and unit cell data are listed in Table 14. The infrared spectrum of solid $NbOCl_3$ contains a strong band at 770 cm^{-1} which is attributed to the Nb–O–Nb stretching vibration, with metal–chlorine frequencies at 429, 405 and 295 cm^{-1}. However, it was recently reported[135] that the Raman spectrum of gaseous $NbOCl_3$ is very similar to that of $VOCl_3$ with a metal–oxygen vibration at 997 cm^{-1} (ν_2), presumably Nb=O, and metal–halogen vibrations at 448 (ν_4), 395 (ν_1), 225 (ν_5), 110 (ν_6) and 106 (ν_3) cm^{-1}. On the basis of this evidence it would appear that $NbOCl_3$ exists in the gaseous state as a monomeric molecule possessing pyramidal (C_{3v}) symmetry. Structural data are not available for the other oxytrihalides, but on the basis of infrared spectroscopy solid $NbOBr_3$, in contrast to $PaOBr_3$ [98], contains similar bridging oxygen atoms to $NbOCl_3$.

The oxytrihalides are moisture-sensitive solids which, apart from $NbOI_3$, decompose with the formation of the pentahalide at high temperatures, although $NbOCl_3$ can be sublimed unchanged in a vacuum at about 200° (Table 7). The tantalum compounds are less stable than their niobium analogues; $TaOCl_3$ decomposes via the dioxymonochloride, TaO_2Cl, at 300°, this product subsequently disproportioning to the pentoxide and pentachloride at 540°. This behaviour is analogous to that of the pentavalent protactinium oxybromides and oxyiodides[75, 98]. Various thermodynamic properties are listed in Tables 6–8 inclusive.

[134] D. E. Sands, A. Zalkin and R. E. Elson, *Acta Cryst.* **12** (1959) 21.
[135] G. A. Ozin and D. J. Reynolds, *Chem. Commun.* (1969) 884.

The electrophilic character of the oxytrichlorides is much less than that of the pentachlorides. Oxyhalide complexes obtained by reactions involving the pentahalides have been mentioned above. Others such as $NbOX_3.L$ (X = Cl and Br; L = diarsine, 2,2′-dipyridyl and *o*-phenanthroline) and $NbOCl_3.L$ (L = for example, alcohols, β-diketones, N,N-diethylaniline, triethylamine and 2-methylquinone) are prepared by direct reaction between the ligand and the appropriate oxytrihalide. For the former series of complexes infrared results indicate the presence of metal–oxygen double bonds.

Dioxymonohalides of the type MO_2X (X = F, Cl and I) are known[72–74] for both niobium and tantalum. The dioxymonofluorides, which possess the ReO_3-type structure, are thermally unstable, decomposing to oxyfluorides of the type M_3O_7F. Other oxyfluorides (Table 14) are obtained by fusion of mixtures of NbO_2F and niobium pentoxide. The oxychlorides Nb_3O_7Cl[72–74] and $Ta_2O_3Cl_4$ are also known. The latter, prepared by the reaction between dichlorine monoxide and $TaCl_5$ in carbon tetrachloride, is analogous to $Pa_2O_3Cl_4$[98].

Little is known about the chemistry of these oxyhalides, but it is interesting to note that the MO_2^+ vibrations in the infrared spectra of NbO_2F and TaO_2F are found around 1000 cm^{-1}, whereas for various actinide compounds containing this group they occur around 800 cm^{-1}. Available crystallographic data are given in Table 14, where references to structural studies are cited.

Halogeno- and Oxyhalogeno-complexes

Alkali metal pentachloroniobates(III) and tantalates(III) of the type $M^I_2M^{III}Cl_5$ (M^I = Na, K, Rb and Cs; M^{III} = Nb and Ta) are obtained by high-temperature reactions between the component halides. These are red, high melting solids which are unstable towards moisture. Nothing is known concerning their chemical properties. Analogous trivalent bromo- and iodo-complexes are unknown.

Tetravalent hexachloro-complexes of the type $M^I_2M^{IV}Cl_6$ (M^I = alkali metal; M^{IV} = Nb and Ta) are obtained by reactions between the appropriate halides either in hydrochloric acid or at high temperature in the absence of air, and by reduction of the corresponding hexachlorometallate(V) with red phosphorus. More recently[136] the caesium salts $Cs_2M^{IV}X_6$ (M^{IV} = Nb and Ta; X = Cl, Br and I) have been prepared by heating CsI with the hexahalogenometallate(V) (CsM^VX_6) or by refluxing similar mixtures in hexachlorobutadiene (chloro-complexes only). The hexachloro-salts are high melting solids which are oxidized in air and water but are moderately stable in hydrochloric acid solutions. They crystallize with cubic or tetragonal symmetry. The caesium hexahalogeno-salts, like their tungsten(IV) analogues, possess face-centred cubic symmetry (K_2PtCl_6-type of structure) and they are paramagnetic. Metal–halogen stretching vibrations occur at 314 and 165 cm^{-1} (Cs_2NbCl_6), 297 and 160 cm^{-1} (Cs_2TaCl_6), 236 and 112 cm^{-1} (Cs_2NbBr_6) and 217 and 109 cm^{-1} (Cs_2TaBr_6). There is some evidence for the existence of the $NbCl_5^-$ ion in the CsCl–$NbCl_4$ phase system, but complexes containing this anion have not been isolated.

$(NEt_4)_2NbCl_6$ and $(NEt_4)_2NbBr_6$ have been prepared[103] by direct interaction of the component halides in chloroform/methyl cyanide mixtures, in a similar manner to the analogous tetravalent actinide hexahalogeno-complexes[98]. Nb–Cl and Nb–Br stretching vibrations (ν_3) are observed at 305 and 222 cm^{-1}, respectively. Other tetravalent niobium

[136] S. M. Horner, R. J. H. Clarke, B. Crociani, D. B. Copley, W. W. Horner, F. N. Collier and S. Y. Tyree (Jr.), *Inorg. Chem.* 7 (1968) 1859.

chloro-complexes known at present include those of the type $(BH)_2Nb(OR)Cl_5$ (R = Me, Et and i-Pr; B = a variety of amines, pyridine and quinoline). These paramagnetic complexes are monomeric in solution.

The niobium(IV) oxytetrachloro-complexes Rb_2NbOCl_4 and Cs_2NbOCl_4 are also known and some weakly paramagnetic dioxyfluoro-complexes of the type M^INbO_2F (M^I = Li, Na and K) have been prepared by heating niobium dioxide with the appropriate alkali metal fluoride in a hydrogen atmosphere.

Both niobium and tantalum form numerous pentavalent fluoro- and oxyfluoro-complexes, many of which can be isolated from aqueous solution. However, only oxychloro- and oxybromo-complexes can be prepared in aqueous media, the preparation of the fully-halogenated compounds requiring the use of non-aqueous solvents. In aqueous fluoride media there exists a distribution of the species $M^VF_n^{5-n}$ ($n < 8$) together with related oxy- or hydroxyanions. Predominant species in specific solutions, as deduced from Raman and n.m.r. spectra, are $NbOF_5^{2-}$ (0–35% HF), NbF_6^- (25–50% HF), TaF_6^- (24 M HF, no TaF_7^{2-}), TaF_7^{2-} (5·2 M NH_4F, no TaF_6^-) and $TaF_6^-+TaF_7^{2-}$ (3·1 → 11·0 M HF). The NbF_6^- anion has been shown to be present in ethanolic solutions of NbF_5 by n.m.r. studies. Presumably this is due to self-ionization. Formation constants (25°) for TaF_6^-, TaF_7^{2-} and TaF_8^{3-} are $4{\cdot}6\times10^3$, $1{\cdot}26\times10^3$ and 4·5, respectively. The NbF_7^{2-} and TaF_8^{3-} anions have not been detected by Raman techniques in hydrofluoric acid solutions up to 12 M and 24 M, respectively. There is conflicting evidence for the existence of TaF_9^{4-} in aqueous fluoride media and no solid complexes containing this anion have yet been prepared. Spectrophotometric and ion exchange studies have indicated that the species $Nb(OH)_2Cl_4^-$ (e.g. 11–13·4 M HCl), $Nb(OH)Cl_3^+$ (e.g. 10·5 M H^+; 0·025 M Cl^-) and $Nb(OH)_2Cl_3$ (~3 M HCl) are present in aqueous hydrochloric acid solutions. However, Raman studies by different groups are reported to support the presence of the $NbCl_6^-$ and $NbOCl_5^{2-}$ ions in concentrated hydrochloric[137].

Octafluoro-, heptafluoro- and hexafluoro-metallates(V) of the types $M_3^IM^VF_8$ (M^I = Na, K and NH_4), $M_2^IM^VF_7$ (M = Na, K and NH_4) and $M^IM^VF_6$ (M^I = Li, Na, K, Rb, Cs, NH_4, Ag, Tl and NO) are prepared either by direct union of the component fluorides at high temperature, or by reactions in aqueous hydrofluoric acid or bromine trifluoride[72–74]. The last method is particularly useful for hexafluoro-complexes which only crystallize from aqueous solution at very high concentrations of hydrofluoric acid.

During the characterization of the I_2^+ cation it was recently demonstrated that iodine reacts with both niobium and tantalum pentafluoride in iodine pentafluoride to form blue solutions.[138]. I_2TaF_6 was isolated from solution, but the niobium(V) complex was found to be unstable in the solid state.

By variations in the reaction conditions oxyfluoroniobates(V) such as K_3NbOF_6, $K_3NbOF_6.H_2O$, $K_2NbOF_5.H_2O$ and $K_2NbOF_5.KHF_2$ are obtained from aqueous hydrofluoric acid, but the preparation of oxyfluorotantalates(V) from aqueous solution is more difficult. The commercial importance of K_2TaF_7 and $K_2NbOF_5.H_2O$ has already been discussed (page 555). The latter loses its water of crystallization at 100° in a vacuum. Ammonium and potassium oxyhexafluorometallates(V), $M_3^IM^VOF_6$ (M^V = Nb and Ta), are conveniently prepared by reactions in methanol and numerous miscellaneous oxyfluoro-complexes, for example, $Na_2Nb_2O_5F_2$, K_2NbO_3F, $M^IM_2^VO_5F$ (M^I = Na and K) and

[137] J. E. D. Davies and D. A. Long, *J. Chem. Soc.* (A) (1968) 2560.
[138] R. D. W. Kemmitt, M. Murray, V. M. McRae, R. D. Peacock, M. C. R. Symons and T. A. O'Donnell, *J. Chem. Soc.* (A) (1968) 862.

$M^{I}Nb_6O_{15}F$ (M^{I} = Li and Na), are formed by fusion of mixtures of the pentoxide and the appropriate alkali metal fluoride[73]. By the addition of hydrogen peroxide to aqueous solutions of oxyfluoro-complexes it is possible to obtain peroxyfluoro-complexes such as $K_2M^{V}O_2F_5.H_2O$, $Rb_2M^{V}O_2F_5.H_2O$ and $Na_3NbO_2F_6.H_2O$.

TABLE 18. CRYSTALLOGRAPHIC DATA FOR SOME PENTAVALENT FLUORO- AND OXYFLUORO-COMPLEXES†

Compound	Symmetry	Space group or structure type	Lattice parameters					Reference
			a_0 (Å)	b_0 (Å)	c_0 (Å)	α_0 (°)	β (°)	
*Na_3TaF_8	Monoclinic	C^6_{2h}–$C2/c$	11·52	5·38	11·21	—	120·9	a
*K_2NbF_7	Monoclinic	C^5_{2h}–$P2_1/c$	5·85	12·67	5·80	—	90·0	b; c
$LiNbF_6$	Rhombohedral	$LiSbF_6$	5·47	—	—	58·09	—	d
$NaNbF_6$	Cubic	$NaSbF_6$	8·28	—	—	—	—	d
*$KNbF_6$	Tetragonal	D^6_{2d}–$P\bar{4}c2$	5·18	—	10·05	—	—	e
$AgNbF_6$	Tetragonal	$KNbF_6$	4·968	—	9·551	—	—	d
$TlNbF_6$	Rhombohedral	$KOsF_6$	5·142	—	—	96·37	—	d
$RbNbF_6$	Rhombohedral	$KOsF_6$	5·14	—	—	96·4	—	d
NH_4NbF_6	Rhombohedral	$KOsF_6$	5·19	—	—	96·1	—	e
$CsNbF_6$	Rhombohedral	$KOsF_6$	5·32	—	—	95·8	—	d
$NONbF_6$	Cubic	—	10·22	—	—	—	—	f
K_3NbOF_6	Cubic	O^5_h–$Fm3m$	8·88	—	—	—	—	g
$(NH_4)_3NbOF_6$	Cubic	O^5_h–$Fm3m$	9·31	—	—	—	—	g
*K_2NbOF_5	Tetragonal	D^{17}_{4h}–$P4/nmm$	6·12	—	8·98	—	—	h
*$NaNbO_2F_2$	Monoclinic	C^5_{2h}–$P2_1/c$	8·063	5·405	7·626	—	101·7	o
KNb_2O_5F	Tetragonal	D^5_{4h}–$P4/mbm$	12·632	—	3·950	—	—	i
$NaNb_2O_5F$	Tetragonal	D^5_{4h}–$P4/mbm$	12·355	—	3·943	—	—	m
*K_2NbO_3F	Tetragonal	K_2NiF_6	3·956	—	13·670	—	—	j
$Na_2Nb_2O_5F_2$	Monoclinic	C^6_{2h}–$C2/c$ or C^4_s–Cc	12·91	7·45	18·24	—	90	m
*$LiNb_6O_{15}F$	Orthorhombic	D^{16}_{2h}–$Pnma$	16·64	3·96	8·89	—	—	k
*$NaNb_6O_{15}F$	Orthorhombic	C^{14}_{2v}–$Amm2$	3·949	10·192	14·721	—	—	l
*$MoNb_{15}O_{40}F$	Monoclinic	C^3_2–$C2$	22·29	3·831	20·27	—	114	n

† Apart from the last 6 niobium compounds listed analogous, isostructural, virtually isodimensional tantalum(V) complexes are known.

* Structural data are given in the references quoted.

[a] J. L. Hoard, W. J. Martin, M. E. Smith and J. F. Whitney, *J. Am. Chem. Soc.* **76** (1954) 3820.
[b] J. L. Hoard, *J. Am. Chem. Soc.* **61** (1939) 1252.
[c] G. M. Brown and A. L. Walker, *Acta Cryst.* **20** (1966) 220.
[d] R. D. W. Kemmitt, D. R. Russell and D. W. A. Sharp, *J. Chem. Soc.* (1963) 4408.
[e] H. Bode and H. von Dohren, *Acta Cryst.* **11** (1958) 80.
[f] N. K. Jha, Ph.D. Thesis, University of British Columbia (1965).
[g] A. E. Baker and H. M. Haendler, *Inorg. Chem.* **1** (1962) 127.
[h] Z. G. Pinsker, *Soviet Physics—Crystallography*, **11** (1967) 634.
[i] A. Magneli and S. Nord, *Acta Chem. Scand.* **19** (1965) 1510.
[j] F. Galasso and W. Darby, *J. Phys. Chem.* **66** (1962) 1318.
[k] M. Lundberg, *Acta Chem. Scand.* **19** (1965) 2274.
[l] S. Andersson, *Acta Chem. Scand.* **19** (1965) 2285.
[m] S. Andersson, *Acta Chem. Scand.* **21** (1967) 1777.
[n] J. Galy and S. Andersson, *Acta Cryst.* **24B** (1968) 1027.
[o] S. Andersson and J. Galy, *Acta Cryst.* **25B** (1969) 847.

Available crystallographic data for the anhydrous complexes are listed in Table 18. Na_3TaF_8 possesses an eight-coordinate structure with the fluorine atoms at a mean distance of 1.98 Å. The symmetry of the TaF_8^{3-} ion is almost that of an ideal square antiprism[139] (D_{4d}). In contrast to this, cubic eight-coordination is observed[140] for the actinide(V) octafluoro-complexes of the type $Na_3M^VF_8$ (M^V = Pa, U and Np). K_2NbF_7 and K_2TaF_7 contain seven-coordinate metal ions, six fluorine atoms being arranged as a distorted trigonal prism with the seventh bonded through one of the faces of the distorted prism[141] (Fig. 6). The symmetry of the Nb_7^{2-} ion approximates to C_{2v}. Although Rb_2UF_7, Rb_2NpF_7 and Rb_2PuF_7 crystallize with this structure all the known heptafluoro-protactinates(V) (M^I = K, Rb, Cs and NH_4) possess an entirely different, nine-coordinate structure[98], presumably as a consequence of the larger ionic radius of Pa(V) relative to those of niobium(V), tantalum(V) and the other pentavalent actinide ions.

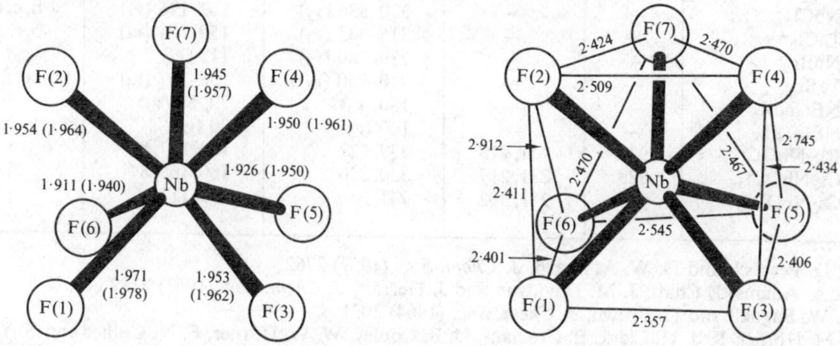

FIG. 6. The coordination polyhedron in K_2NbF_7 [141].
Values in parentheses have been corrected for thermal motion.
(Reproduced by permission of the copyright holder.)

The lattice of $KNbF_6$ contains discrete NbF_6^- ions[142], the six fluorine atoms being arranged as a slightly distorted octahedron with a mean Nb–F bond distance of 2·14 Å. $KTaF_6$, $AgNbF_6$, $AgTaF_6$ and many of their *d*-transition metal analogues crystallize with this arrangement[74], but the hexafluoro-salts with larger cations (Tl, Rb, Cs and NH_4) possess the rhombohedral six-coordinate $KOsF_6$-type of structure (Table 18) in common with their *d*-transition metal analogues[74] and the caesium salts of uranium(V), neptunium(V) and plutonium(V)[98]. In contrast to this, other pentavalent actinide hexafluoro-complexes (M^I = K, Rb, Cs and NH_4) are isostructural with $RbPaF_6$ which is eight-coordinate with four terminal and four bridging fluorine atoms[98].

Both $K_2NbOF_5.KHF_2$, which is monoclinic with a_0 = 8·82, b_0 = 14·02, c_0 = 6·82 Å and β = 86·6°, space group C_{2h}^6–$C2/c$, and K_2NbOF_5 (Table 18) contain discrete $NbOF_5^{2-}$ ions. In the latter[143] this group is bipyramidal with C_{4v} symmetry, the niobium–oxygen bond length being 1·68 Å ($\upsilon_{Nb=O}$ at 926 cm^{-1}) with niobium–fluorine distances of 2·06 Å (Nb–F_1) and 1·84 Å (Nb–4F_2). The metal–oxygen stretching vibrations for the above complexes occur around 890 cm^{-1} (Ta) and 930 cm^{-1} (Nb). Metal–fluorine vibrations are

139 J. L. Hoard, W. J. Martin, M. E. Smith and J. F. Whitney, *J. Am. Chem. Soc.* **76** (1954) 3820.
140 D. Brown, J. F. Easey and C. E. F. Rickard, *J. Chem. Soc.* (A) (1969) 1161.
141 G. M. Brown and L. A. Walker, *Acta Cryst.* **20** (1966) 220.
142 H. Bode and H. von Dohren, *Acta Cryst.* **11** (1958) 80.
143 G. Z. Pinsker, *Soviet Physics Cryst.* **11** (1966) 634.

found between 630 and 260 cm^{-1} in their infrared spectra. Raman spectral results are available[144, 145] for solid Na_3TaF_8, K_2NbF_7, K_2TaF_7, $CsNbF_6$, $K_2NbOF_5.H_2O$, K_3TaOF_6 and $(NH_4)_3NbOF_6$ and for various anionic species present in aqueous fluoride media; some infrared data are listed in Table 19.

TABLE 19. INFRARED DATA FOR NIOBIUM(V) AND TANTALUM(V) HALOGENO- AND OXYHALOGENO-COMPLEXES

Compound/Anion	$\nu_{M=O}$	ν_{M-O} rock	ν_{M-X} stretch	ν_{X-M-X} def.	Reference
NbF_7^{2-}	—	—	524	—	a
TaF_7^{2-}	—	—	518	—	a
MF_6^-	—	—	580	—	a
$NbCl_6^-$	—	—	330–336 (ν_3)	162–176 (ν_4)	b, c, d, e
$TaCl_6^-$	—	—	315–333 (ν_3)	150–164 (ν_4)	b, c, d, e
$NbBr_6^-$	—	—	239–240 (ν_3)	112 (ν_4)	e, f, g
$TaBr_6^-$	—	—	210–230 (ν_3)	106–107 (ν_4)	e, f, g
NEt_4NbI_6	—	—	180 (ν_3)	70, 66 (ν_4)	h
NEt_4TaI_6	—	—	160 (ν_3)	80 (ν_4)	h
Rb_2NbOCl_5	930	231, 219	339, 327	180, 167, 87	i
Cs_2NbOCl_5	928	231, 217	330, 320	180, 168, 87	i
Cs_2NbOBr_5	977	211, 202	241	140, 111	i

[a] R. D. Peacock and D. W. A. Sharp, *J. Chem. Soc.* (1959) 2762.
[b] D. A. Adams, J. Chatt, J. M. Davidson and J. Gerratt, *J. Chem. Soc.* (1963) 2189.
[c] K. W. Bagnall and D. Brown, *J. Chem. Soc.* (1964) 3021.
[d] S. M. Horner, R. J. H. Clark, B. Crociani, D. B. Copley, W. W. Horner, F. N. Collier and S. Y. Tyree (Jr.), *Inorg. Chem.* **7** (1968) 1859.
[e] W. van Bronswyk, R. J. H. Clark and L. Maresca, *Inorg. Chem.* **8** (1969) 1395.
[f] D. Brown and P. J. Jones, *J. Chem. Soc.* (A) (1967) 247.
[g] I. R. Beattie, T. R. Gilson and D. A. Ozin, *J. Chem. Soc.* (A) (1968) 2765.
[h] G. A. Ozin, G. W. A. Fowles, D. J. Tidmarsh and R. A. Walton, *J. Chem. Soc.* (A) (1969) 642.
[i] A. Sabatini and I. Bertini, *Inorg. Chem.* **5** (1966) 204.

Anhydrous hexachloro-niobates(V) (yellow) and tantalates(V) (white) of the type $M^IM^VCl_6$ (M^I = alkali metal, NEt_4, NMe_4, Ph_4As, etc.) are only obtained by reactions in non-aqueous solvents such as thionyl chloride or iodine monochloride, or by high-temperature reactions between the component chlorides[72–74]. Hydrated complexes are unknown. Although niobium(V) and tantalum(V) do not form octachloro-complexes with univalent cations analogous to those known for pentavalent protactinium and uranium it is reported that heptachloro- and octachloro-complexes of the types $M^{II}M^VCl_7$ (M^{II} = Ba, Sr, Ca and Pb) and GaM^VCl_8 (M^V = Nb and Ta) can be prepared by direct union of the component halides at high temperatures[146]. Hexabromo-complexes (M^I = NEt_4 and NMe_4) have been prepared using methyl cyanide or thionyl bromide as solvents[74], and a recent publication[147] describes the preparation of NEt_4NbI_6 and NEt_4TaI_6 using chloroform/methyl cyanide mixtures as the solvent. Mixed hexahalogeno-complexes of the type $NEt_4MX_5.Y$ (X = Cl or Br; Y = Cl, Br or I) are also obtained by reactions in this solvent.

144 O. L. Keller (Jr.), *Inorg. Chem.* **2** (1963) 783.
145 O. L. Keller (Jr.) and A. Chetham-Strode (Jr.), *Inorg. Chem.* **5** (1966) 362.
146 N. D. Chikanov, *Zhur. Neorg. Khim.* **13** (1968) 1191 and 2616.
147 G. A. Ozin, G. W. A. Fowles, D. J. Tidmarsh and R. A. Walton, *J. Chem. Soc.* (A) (1969) 642.

Reactions in aqueous hydrochloric acid or hydrobromic acid yield[72–74] oxypentahalogenoniobates(V) of the type $M^I_2NbOX_5$ (X = Cl or Br; M^I = variously, K, Rb, Cs, pyridinium, etc.) but owing to the greater tendency of tantalum(V) to undergo hydrolysis in aqueous media no well-defined oxyhalogenotantalates(V) can be obtained from aqueous solution. However, like their niobium(V) analogues, oxypentachloro- and oxytetrachlorotantalates(V), $M^I_2TaOCl_5$ and M^ITaOCl_4 respectively (M^I = alkali-metal), are obtained by heating the appropriate hexachlorometallate(V) with the stoichiometric amount of Sb_2O_3 either in chlorine or, better, in a vacuum. Alternatively, the niobium(V) complexes are obtained by heating together $NbOCl_3$ and the appropriate alkali–metal chloride. $NONbOCl_4$ is prepared by the reaction between nitrosyl chloride and niobium oxytrichloride. $(C_5H_5N)NbOCl_4$ and $(C_9H_7NH)NbOCl_4$ have been prepared from aqueous hydrochloric acid solution and on the basis of infrared data it is suggested[148] that they contain Nb–O–Nb–O chains (υ_{Nb-O} at 850 and 800 cm^{-1}). A few peroxy derivatives of the type $M^IM^V(O_2)Cl_5$ (M^I = K, Rb, Cs and NH_4) are also known.

Although X-ray powder results are available for many of the above complexes, the only structural information is that for Cs_2NbOCl_5 which, in common with other oxypentachloroniobates(V) and Cs_2NbOBr_5, possesses the K_2PtCl_6-type structure[149].

The hexachloro-complexes are all moisture-sensitive, thermally unstable solids (Table 8) for which the heats of formation increase with the size of the alkali–metal cation. For example, ΔH_f for $LiNbCl_6$ and $CsNbCl_6$ is $-1{\cdot}9$ and $-24{\cdot}4$ kcal $mole^{-1}$, respectively. The oxychloro-complexes are also thermally unstable, those of niobium(V) evolving $NbOCl_3$ at high temperature (Table 8), whereas those of tantalum(V) evolve tantalum pentachloride. This difference is probably a consequence of the relative thermal instability of $TaOCl_3$.

Some characteristic infrared stretching vibrations are listed in Table 19. The Raman spectra of certain solid hexachloro- and hexabromo-complexes[124, 150] and of several mixed hexahalogeno-complexes[147] have also been reported and results for nitromethane and methyl cyanide solutions of NEt_4TaCl_6 and NEt_4NbCl_6, respectively, indicate that the $TaCl_6^-$ and $NbCl_6^-$ ions are the predominant species in such solutions. Force constants have been calculated for the hexahalogeno species[150]. In contrast to the oxypentachloro-complexes, peroxypentachloro-complexes show characteristic double peaks in the infrared regions 860–900 and 565–600 cm^{-1}.

2.5. CYANIDES, THIOCYANATES AND SELENOCYANATES

Binary cyanides of niobium and tantalum are unknown, but the monosubstituted compound $NbCl_4.CN(C_2H_5)_2O$ is obtained as reddish-brown crystals when anhydrous hydrogen cyanide is passed into a slurry of the pentachloride in ether. In the absence of solvent, $H(NbCl_5.CN)$ is obtained[151]. The analogous bromo-compound is also known. Little is known of the properties of these compounds and no cyano-complexes have been characterized.

Tetravalent niobium hexathiocyanato-complexes of the type $M^I_2Nb(NCS)_6$ (M^I = K and Ph_4As) are obtained from the tetrachloride by metathesis with the appropriate thiocyanate in methyl cyanide[152].

148 I. S. Morozov and N. P. Lipatova, *Russ. J. Inorg. Chem.* **13** (1968) 1101.
149 D. Brown, *J. Chem. Soc.* (1964) 4944.
150 W. van Bronswyk, R. J. H. Clark and L. Maresca, *Inorg. Chem.* **8** (1969) 1395.
151 G. Brauer and H. Walz, *Z. anorg. und allgem. Chem.* **319** (1963) 236.
152 T. M. Brown and G. F. Knox, *J. Am. Chem. Soc.* **89** (1967) 5296; *Inorg. Chem.* **8** (1969) 1401.

Solvated pentavalent thiocyanates of the types $[M^V(NCS)_5.CH_3CN].2CH_3CN$ and $M^IM^V(NCS)_6.xCH_3CN$ (M^I = alkali metals, NH_4 and PyH; x = variously 2, 5 or 7; M^V = Nb and Ta) are obtained by similar metatheses in anhydrous methyl cyanide. The methyl cyanide may be removed by vacuum heating, the anhydrous compounds being obtained at 80–115°[152–155]. Similar reaction in ether result in the formation of mixed chloro-thiocyanato-complexes of the type $M^VCl_a(NCS)_b.(C_2H_5)_2O$ (a = 3 or 4, b = 1 or 2); methyl cyanide adducts with a and $b = 1 \rightarrow 4$ are also known. The thiocyanate groups appears to be bonded via the nitrogen atom in these compounds and it is suggested that the pentathiocyanates are dimeric.

The heptathiocyanato-complex, $K_2Nb(NCS)_7.3DMF$, which is a pink, air-sensitive solid, and the mixed complex $KNb(NCS)_4(OMe)_2.CH_3OH$ are prepared by metatheses in dimethylformamide (DMF) and methanol, respectively[156].

It is claimed[157] that the oxythiocyanate complexes $NbO(SCN)_3.C_6H_{10}O$ and $NbO(SCN)_3.C_9H_{14}O$ are obtained when niobium pentachloride is reacted with potassium thiocyanate in anhydrous acetone, the mesityl oxide and phorone resulting from the condensation of acetone under these conditions. In the presence of controlled amounts of water this reaction is reported to lead to the formation of NbO_2SCN.

Pentavalent selenocyanates of the type $KM^V(NCSe)_6$ (M^V = Nb and Ta) are also obtained by metatheses in anhydrous methyl cyanide[158]. The selenocyanate groups also appear to be coordinated via the nitrogen atom, but these complexes are less stable than the thiocyanates, liberating selenium on contact with water.

2.6. OXIDES AND ANHYDROUS MIXED OXIDES

In addition to the stoichiometric oxides discussed below, several sub-oxide phases such as M_6O, M_4O, Nb_2O and TaO have been identified crystallographically. However, sesquioxides analogous to V_2O_3 are unknown. The kinetics and conditions of formation of the sub-oxides have been extensively studied[73, 159] in view of their importance in connection with the resistance of the metals to high-temperature oxidation and the effect of oxygen on the mechanical properties of the metals. The preparation and crystallographic properties of the various tantalum sub-oxides and the evidence for the existence of TaO and TaO_2 are reviewed in detail by Niebuhr[159] whilst others describe recent studies on the niobium–oxygen system[160].

Niobium Monoxide

Hydrogen reduction of Nb_2O_5 (1300–1700°) or niobium reduction of NbO_2 in argon (1700°) results in the formation of NbO, which is a stoichiometric oxide with a range of homogeneity extending at the most from $Nb_{1.04}$ to $NbO_{0.94}$. It possesses cubic symmetry (Table 20) and forms cubic mixed crystals with vanadium monoxide. The enthalpy and free energy of formation of NbO are, respectively, −97·2 and −90·5 kcal $mole^{-1}$. Tantalum monoxide has not been prepared.

153 H. Böhland and E. Zenker, *J. Less-Common Metals*, **14** (1968) 397.
154 H. Böhland and E. Tiede, *J. Less-Common Metals*, **13** (1967) 224.
155 H. Böhland, E. Tiede and E. Zenker, *J. Less-Common Metals*, **15** (1968) 89.
156 A. M. Goluls and A. M. Sych, *Russ. J. Inorg. Chem.* **9** (1964) 593.
157 G. B. Seifer, Yu. Ya. Kharitonov and Z. A. Tarasova, *Inorg. Mat.* **4** (1968) 444.
158 B. L. Bush and T. M. Brown, Abstract 157th Meeting Amer. Chem. Soc., April 1969, No. 118.
159 J. Niebuhr, *J. Less-Common Metals*, **10** (1966) 312; ibid. **11** (1966) 191.
160 A. Taylor and N. J. Doyle, *J. Less-Common Metals*, **13** (1967) 313.

Niobium Dioxide

Reduction of Nb_2O_5 with hydrogen at 800–1350° or with niobium metal in an electric arc furnace leads to the formation of NbO_2. Alternatively, it may be obtained by the thermal decomposition of the pentoxide at 1150° in argon. It is a bluish-black solid which reduces sodium hydroxide, carbon dioxide and sulphur dioxide at moderate temperatures. It is, however, insoluble in water.

Crystal structure studies (Table 20) have shown that NbO_6 octahedra are coupled together by common edges to form strings with successive Nb–Nb bond distances of 2·80 and 3·20 Å (cf. α-NbI_4 [100]) with associated metal–oxygen distances to the common edge of 3·0 and 2·6 Å respectively. Like the tetrahalides NbO_2 is diamagnetic. Thermodynamic functions are listed in Table 21.

Although TaO_2, like TaO, has been observed mass-spectrographically and homogeneous mixed crystals with NbO_2 (0·8 NbO_2, 0·2 TaO_2 to 0·4 NbO_2, 0·6 TaO_2) can be prepared there is no clear evidence for its existence as an independent compound.

TABLE 20. CRYSTALLOGRAPHIC DATA FOR CERTAIN OXIDE PHASES*

Phase†	Symmetry	Space group	Lattice parameters				$\rho_{calc.}$ g cm^{-3}	Reference
			a_0 (Å)	b_0 (Å)	c_0 (Å)	β (°)		
NbO	Cubic	O_h^1–$Pm3m$	4·2108	—	—	—	7·30	a
NbO_2	Tetragonal	C_{4h}^6–$I4_1/a$	13·70	—	5·987	—	5·90	b
B-Nb_2O_5 (ζ, γ^1)	Monoclinic	C_{2h}^6–$P2/b$	7·348	5·962	13·646	115·5	5·29	c
H-Nb_2O_5 (α, β^1)	Monoclinic	C_2^1–$P2$	21·16	3·822	19·35	119·83	4·55	d, e
P-Nb_2O_5 (η)	Tetragonal	—	3·896	—	25·43	—	4·54	f
β-Ta_2O_5	Orthorhombic	C_{2v}^8–Pba	6·20	3·66	3·89	—	—	g

* Only those pentoxide phases for which structural details are available are listed. Structural details are given in references quoted.
† Symbols in parentheses are alternative notations used by different authors.
[a] S. Andersson and A. Magneli, *Acta Chem. Scand.* **11** (1957) 1065.
[b] B.-O. Marinder, *Acta Chem. Scand.* **15** (1961) 707; *Arkiv Khemi*, **19** (1962) 435.
[c] F. Laves, W. Petter and H. Wulf, *Naturwissen.* **51** (1964) 633.
[d] B. M. Gatehouse and A. D. Wadsley, *Acta Cryst.* **17** (1964) 1545.
[e] R. Norin, M. Carlsson and B. Elgquist, *Acta Chem. Scand.* **20** (1966) 2892.
[f] F. Laves and W. Petter, according to ref. 161.
[g] K. Lehovec, *J. Less-Common Metals*, **7** (1964) 397.

Niobium and Tantalum Pentoxide

The ultimate products of the oxidation of the metals in oxygen or air at high temperature are the pentoxides. Alternatively, they are obtained by air ignition of various other compounds, in particular, the hydrous oxides precipitated from aqueous solution. They are white, air-stable, water-insoluble solids which are important starting materials for metal production (page 556). They are not readily soluble in aqueous mineral acids, with the exception of hydrofluoric acid, but following fusion with, for example, alkali metal pyrosulphates, potassium hydroxide or potassium carbonate the resulting melt may be dissolved in acid solution or, depending on its composition, in water.

TABLE 21. THERMODYNAMIC FUNCTIONS FOR NIOBIUM[a] AND TANTALUM[b] OXIDES, CARBIDES, NITRIDES AND BORIDES

Compound	ΔH_f kcal mole^{-1}	ΔG_f (298°K)	S_{298} e.u.	C_p cal mole^{-1} deg^{-1} (298°K)	Heat capacity cal mole^{-1} deg^{-1} $C_p = a+bT+cT^{-2}$			
					a	$b\times10^3$	$c\times10^{-5}$	Temp. range
NbO	−97·2	−90·5	(11·5)*	9·86	10·04	2·35	−0·783	298–1700
NbO_2	−189·7	−177·5	(13·0)	13·79	14·681	6·156	−2·421	298–1040
NbO_2	—	—	—	—	21·28	—	—	1040–1700
Nb_2O_5	−453·5	−421·8	32·8	31·58	38·76	3·54	−7·318	298–1780
$Nb_2O_{5(l)}$	—	—	—	—	57·90	—	—	1780–1810
Ta_2O_5	−489·0	−456·4	34·2	—	—	—	—	—
NbN	−55·9	—	—	10·30	8·69	5·40	—	298–600
Nb_2N	−59·4	—	—	—	—	—	—	—
TaN	−64·7	—	—	—	—	—	—	—
Ta_2N	−59·0	−51·3	—	—	—	—	—	—
NbC	−33·0	−31·2	8·38	8·88	10·79	1·726	−2·15	298–1800
Nb_2C	−44·5		15·3	7·24	7·94	1·50	−1·025	298–1800
TaC	−35·0	−34·7	10·1	—	11·46	1·33	—	1296–2843
Ta_2C	−47·2	—	(19)	—	—	—	—	—
NbB_2	−41·9	—	(8·95)	(11·81)	11·01	9·38	−1·78	298–1200
NbB	—	—	8·91	—	—	—	—	—
TaB_2	—	—	—	—	11·39	8·75	—	298–1200
TaB	—	—	—	—	7·82	5·8	—	298–1200

(l) = liquid, unless otherwise indicated data refer to solid phases.
* Values in parentheses are estimated.
[a] V. I. Lavrentev and Ya. I. Gerassimov, *Atomic Energy Rev.*, Special Issue, No. 2 (1968) 7.
[b] *Metals Reference Book*, Ed. C. J. Smithells, 4th ed. Vol. I, Butterworths, London (1967).

The preparation, structures and thermal properties of the various polymorphic forms of Nb_2O_5 have recently been reviewed in detail[161]. The authors, who correlate the various nomenclatures used by different groups for the polymorphs of Nb_2O_5 prefer to use the designations TT-, T-, B-, M-, H-, N- and P-Nb_2O_5 for the well-established forms. The conversion of one form to another on heating in air is illustrated in Fig. 7, from which it is apparent that H-Nb_2O_5 is the stable form in air above 900°. Single crystals of certain of the forms noted on the figure and also of the B-, N- and P-modifications are obtained by chemical transport reactions in the presence of a halogen. Crystallographic data are given in Table 20 for those forms for which structural details are available. Additional modifications termed ε-Nb_2O_5, Nb_2O_5-I-high, Nb_2O_5-II, R-Nb_2O_5 and Nb_2O_5-OX-I to

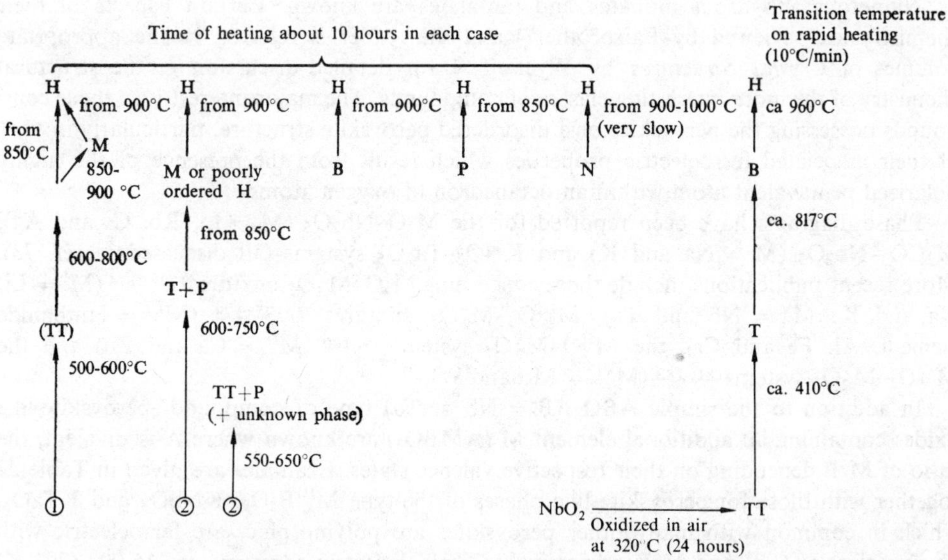

FIG. 7. Interconversion of the various polymorphs of Nb_2O_5 [161]. (Reproduced by permission of the copyright holder.)

Nb_2O_5-OX-VI have also been observed, but further work is necessary to clarify the exact nature of most of these phases. The preparation and structures of all the above modifications are discussed in detail by Schäfer *et al.*[161] and the results of a survey of the properties of non-stoichiometric niobium and tantalum oxide phases have recently been published by Gruehn and Schäfer[162]. Amongst these latter phases structural information is available for $Nb_{12}O_{29}$ [163], $Nb_{22}O_{54}$ [164, 165], $Nb_{25}O_{62}$ [165, 166] and $NbO_{2.483}$ [167], all of which resemble H-Nb_2O_5. It is interesting to note that there is little resemblance between the niobium (or tantalum) oxide system and the protactinium oxide system[168].

161 H. Schäfer, R. Gruehn and F. Schulte, *Angew. Chem. Int. Ed.* **5** (1966) 40.
162 R. Gruehn and H. Schäfer, *J. Less-Common Metals*, **10** (1966) 152.
163 R. Norin, *Acta Chem. Scand.* **17** (1963) 1391; ibid. **20** (1966) 871.
164 R. Norin and A. Magnéli, *Naturwissen.* **47** (1960) 354.
165 R. Norin, M. Carlsson and B. Edgquist, *Acta Chem. Scand.* **20** (1966) 2892.
166 R. Norin, *Naturwissen.* **52** (1965) 300.
167 R. Gruehn and R. Norin, *Z. anorg. und allgem. Chem.* **355** (1967) 176.
168 D. Brown, in *Adv. Inorg. and Rad. Chem.* **12** (1969), page 1, and references therein.

Ta_2O_5 is dimorphic, the transition $\beta \rightarrow \alpha$-Ta_2O_5 occurring only slowly at 1360°. β-Ta_2O_5 is isostructural with N-Nb_2O_5 whilst the α-form is similar to P-Nb_2O_5, possessing tetragonal symmetry with $a_0 = 3{\cdot}795$ and $c_0 = 35{\cdot}54$ Å. The melting points of α-Ta_2O_5 and β-Ta_2O_5 are respectively $1872 \pm 10°$ and $1785 \pm 30°$ (cf. H-Nb_2O_5 which melts at approximately 1500°). Thermodynamic functions for the pentoxides are given in Table 21.

The mixed oxide $2Nb_2O_5.Ta_2O_5$, which is tetragonal with $a_0 = 15{\cdot}77$ and $c_0 = 3{\cdot}84$ Å, is formed at 1400°. It is apparently isostructural with phases of the type $AM^V_9O_{25}$ (A = P, As and V; M^V = Nb and Ta)[169]. Mixed oxides with Pa_2O_5, $M^V_2O_5.3Pa_2O_5$, are also known[170].

Mixed Oxide Phases

Numerous anhydrous niobates and tantalates are known. Certain aspects of their chemistry are reviewed by Fairbrother[73] and the reader is referred to the appropriate volumes of *Crystal Structures* by Wyckoff[171] for detailed discussion of the structural chemistry of the more interesting classes of compounds. The major interest is in those compounds possessing the perovskite or a disordered perovskite structure, particularly in view of their associated ferroelectric properties which result from the presence of the highly polarized pentavalent atom within an octahedron of oxygen atoms.

Phase diagrams have been reported for the M_2O–Nb_2O_5 (M = Li, Rb, Cs and Ag), M_2CO_3–Nb_2O_5 (M = Na and K) and KNO_3–Ta_2O_5 systems (all discussed in ref. 73). More recent publications include those concerning M^I_2O–$M^V_2O_5$ mixtures[172–2174] (M^I = Li, Na and K; M^V = Nb and Ta), $M^{III}_2O_3$–$M^V_2O_5$ mixtures[173, 175–178] (M^{III} = lanthanide element, Al, Fe and Cr), the $M^{II}O$–$M^V_2O_5$ systems[179, 180] (M^{II} = Ca and Zn) and the $M^{VI}O_3$–$M^V_2O_5$ systems[181, 182] (M^{VI} = Mo and W).

In addition to the simple ABO_3 (B = Nb or Ta) type of compound, perovskite-like oxides containing an additional element M ($AMBO_3$) are known where A is divalent, the ratio of M:B depending on their respective valence states. Examples are given in Table 22 together with those for perovskite-like phases of the type $M^{III}_4B_2O_9$. $KNbO_3$ and $KTaO_3$, which in common with many other perovskites are polymorphic, are ferroelectric with Curie points at 688° and 13°K respectively. Their dielectric constants are 15,000 (707°K) and 4700 (13°K), respectively. High-temperature crystallographic studies of the cubic forms of $KNbO_3$ and $KTaO_3$ (480°) indicate that the Ta–O bond is slightly shorter than the Nb–O bond, the lattice parameters of the pure samples being 4·0026 for $KTaO_3$ and 4·0226 Å for $KNbO_3$. $NaNbO_3$, which is cubic at 640°, exists in three other crystal forms at lower temperatures, but there is little difference between the various atomic arrangements[183].

169 J. Waring and S. Roth, *Acta Cryst.* **17** (1964) 455.
170 C. Keller, *J. Inorg. Nucl. Chem.* **27** (1965) 321 and 1233.
171 R. W. G. Wyckoff, *Crystal Structures*, 2nd ed., Interscience, New York, Vols. II (1965), III (1965) and IV (1968).
172 C. D. Whiston and A. J. Smith, *Acta Cryst.* **19** (1965) 169.
173 C. D. Whiston, Ph.D. Thesis, Sheffield University (1963).
174 A. Deschanvres, C. Michel and B. Raveau, *Bull. Soc. Chim. Fr.* (1968) 4805.
175 P. N. Iyer and A. J. Smith, *Acta Cryst.* **23** (1967) 740.
176 N. A. Godina, E. P. Sarchenko and E. K. Keler, *Dokl. Akad. Nauk. SSSR*, **178** (1968) 1324.
177 C. Keller and C. H. Walter, *J. Inorg. Nucl. Chem.* **21** (1965) 1247; ibid. **21** (1965) 1253.
178 V. K. Trimov, L. M. Kovba and Z. Ya. Pol'shchikova, *Russ. J. Inorg. Chem.* **13** (1968) 286.
179 H. Kasper, *Z. anorg. und allgem. Chemie*, **355** (1967) 1.
180 D. A. Reeve, *J. Less-Common Metals*, **17** (1969) 215.
181 R. S. Roth and A. D. Wadsley, *Acta Cryst.* **19** (1965) 26, 32, 38, 42.
182 J. Galy and S. Andersson, *Acta Cryst.* **24B** (1968) 1027.
183 R. W. G. Wyckoff, *Crystal Structures*, 2nd ed., Vol. 2, p. 405, Interscience, New York (1965).

TABLE 22. EXAMPLES OF PEROVSKITE-LIKE NIOBATES AND TANTALATES[a]

Type	M, with B = Nb or Ta
ABO_3	A = Na and K
$Ba(M^{I}_{0\cdot25}, Ta_{0\cdot75})O_3$	Na
$Sr(^{I}_{0\cdot25}, Ta_{0\cdot75})O_3$	Na
$Ba(M^{II}_{0\cdot33}, B_{0\cdot67})O_3$	Mg, Co, Ni, Cu, Zn
$Ca(M^{II}_{0\cdot33}, B_{0\cdot67})O_3$	Ni
$Pb(M^{II}_{0\cdot33}, B_{0\cdot67})O_3$	Mg, Ni
$Ba(M^{III}_{0\cdot5}, B_{0\cdot5})O_3$	Mn, Fe, Co, Ni, Rh, Sc, Y, La–Lu, Pu[b], Am[b]
$Pb(M^{III}_{0\cdot5}, B_{0\cdot5})O_3$	Fe, Sc, La–Lu
$Sr(M^{III}_{0\cdot5}, B_{0\cdot5})O_3$	Cr, Ga, In, Y, Pr, Nd, Sm, Yb, Fe
$Ba(M^{I}, M^{IV}B)O_6$	M^{I} = K, Na; M^{IV} = Ti
$Ba_4B_2O_9$	—
$Mg_4B_2O_9$	—

[a] Apart from those marked b, these examples are taken from refs. 73, 151 and 187.
[b] C. Keller and C. H. Walker, *J. Inorg. Nucl. Chem.* **27** (1965) 1247.

Of the four phases observed for $KNbO_3$, only that of rhombohedral symmetry has an ordered structure[184].

Niobates and tantalates of the types $Nb_2M^{II}O_6$ (M^{II} = Cd, Co, Fe, Mg, Mn, Ni and Zn) and $Ta_2M^{II}O_6$ (M^{II} = Cd, Mn and Zn) possess the orthorhombic columbite [$Nb_2(Fe, Mn)O_6$] type of structure in which each metal atom is surrounded by a nearly regular octahedron of oxygen atoms. However, tapiolite, Ta_2FeO_6, and mossite, $(Ta, Nb)_2FeO_6$, both possess a tetragonal layered-structure which can be considered as a superlattice based on cassiterite, SnO_2. $Sn_2Ta_2O_7$, and certain other oxides of the type $M^{II}_2B^{V}O_7$ (M^{II} = Cd and Hg; B^{V} = Nb and Ta) possess the atopite-type of structure[185], as does pyrochlore, one of the mineral sources of niobium and tantalum, whilst other compounds of the type $M^{II}_2B^{V}O_7$ (M^{II} = Pb, Ba and Sr) exhibit a slightly distorted arrangement of this structure. Rare earth niobates of the type $M^{III}NbO_4$ are isostructural with the monoclinic form of $(Y, Yb)NbO_4$, possessing a distorted scheelite structure in which discrete tetrahedral oxyanions exist, but the scandium compounds, $ScB^{V}O_4$, exhibit a different monoclinic distortion of the scheelite structure.

Other oxide systems which have been studied include series of the types $M^{V}B^{V}_9O_{25}$ (M^{V} = V, P or As) and $M^{II}_4Nb_2O_9$ (M^{II} = Mn, Co, Fe and Mg). In contrast to $Ba_4Nb_2O_9$, etc. (Table 22), the latter compounds have structures similar to that of Cr_2O_3 [186]. Coloured tungsten bronze-like phases are known[73] and in the ferroelectric compound $Ba_6Ti_2Nb_8O_{30}$ the titanium and niobium atoms are statistically distributed throughout the structure.

Crystallographic data for the above-mentioned mixed oxide phases are given in the references quoted and, together with similar data for numerous other niobium mixed oxides, in a very recent extensive compilation[187].

[184] R. Comes and M. Lambert, *Solid State Comm.* **6** (1968) 715.
[185] A. W. Sleight, *Inorg. Chem.* **7** (1968) 1704.
[186] R. W. G. Wyckoff, *Crystal Structures*, 2nd ed., Vol. 2, p. 8, Interscience, New York (1965).
[187] H. Nowotny and K. Seifert, *Atomic Energy Rev.*, Spec. Issue, No. 2 (1968) 71.

Examples of mixed oxides containing tetravalent niobium or tantalum include $Ba_{0\cdot5}(Ta)O_{2\cdot5}$, $Ba_{0\cdot78}(Ge_{0\cdot5}, Ta_{0\cdot5})O_3$, $Ba_{0\cdot5}(Ta_{0\cdot5}, Nb_{0\cdot5})O_{2\cdot5}$, $V_{0\cdot25}Nb_{0\cdot75}O_2$, $Ti_{0\cdot75}Nb_{0\cdot25}O_2$ and $Nb_{0\cdot33}Mo_{0\cdot67}O_2$.

Peroxy Compounds

Orthoperoxoniobates(V) and tantalates(V) of general formula $M^I_3M^VO_8$ (M^I = alkali metal and NH_4^+; M^V = Nb and Ta) precipitate from cold alkaline solutions of soluble niobates or tantalates on the addition of hydrogen peroxide[73]. Alternatively, the ammonia salts are precipitated from nitric acid–hydrogen peroxide mixtures on the addition of aqueous ammonia or from hydrogen peroxide solution on the addition of ethanol and aqueous ammonia[188] at 0°C. The alkali–metal salts are stable in the atmosphere, but $(NH_4)_3NbO_8$ and $(NH_4)_3TaO_8$ gradually liberate active oxygen. Peroxo-ortho-acids $HNbO_4$ (yellow) and $HTaO_4$ (white) are formed on the addition of dilute sulphuric acid to the peroxo-salts and the nature of the niobium and tantalum species in such solutions has been studied[189, 190].

The solid orthoperoxo-complexes all possess tetragonal symmetry and are isostructural with K_3CrO_8 [191].

Unstable peroxo-metaniobates(V), $M^INbO_4.nH_2O_2.mH_2O$ (n = 1·3 to 1·5; m = 1·5 to 3·0, M^I = Na and K), are formed by the action of concentrated hydrogen peroxide solution on the appropriate metaniobate. Peroxo-halogeno and peroxo-oxalato complexes are discussed in sections 2.4 and 2.16 respectively.

2.7. HYDROXIDES AND SOLUBLE NIOBATES(V) AND TANTALATES(V)

Insoluble, polymeric, hydrated pentoxides of variable water content are obtained by a variety of methods including hydrolysis of water-soluble complexes and acidification of niobate(V) or tantalate(V) solutions[73]. Precipitation by ammonia generally yields a more chemically active and more finely divided product. Although these hydrous oxides are commonly called "niobic acid" and "tantalic acid" they possess barely discernible acid functions. They are insoluble in water, apart from colloidal behaviour, and the existence of definite hydroxides or hydrates has never been established. Tantalic acid is appreciably less reactive towards complexing agents than niobic acid, the reactivity of which decreases on "aging", behaviour which also characterizes protactinium(V) hydrous oxide. The hydrous oxides are virtually insoluble in dilute aqueous mineral acids apart from hydrofluoric acid, but dissolve in concentrated sulphuric acid, concentrated hydrochloric acid (niobium only) and in chromic and orthophosphoric acid (niobium only). Niobic acid is also soluble in 15 M selenic acid, but reprecipitates on dilution of the solution to about 7 M and, when freshly precipitated, dissolves in concentrated alkaline solutions. Tantalic acid, however, is insoluble in the latter reagents. Both niobic acid and tantalic acid dissolve in oxalic, tartaric and certain α-hydroxycarboxylic acid solutions and they react with a variety of amines and 1,2-diols, the reactivity of tantalic acid being appreciably less than that of niobic acid. Complexes isolated from these solutions are discussed in detail by Fairbrother[73].

188 K. I. Selezneva and L. A. Nisel'son, *Russ. J. Inorg. Chem.* **13** (1968) 45.
189 V. P. Vasil'ev and G. A. Zaitseva, *Russ. J. Inorg. Chem.* **13** (1968) 42.
190 N. Alder and C. F. Hiskey, *J. Am. Chem. Soc.* **79** (1957) 1827, 1831, 1834.
191 R. W. G. Wyckoff, *Crystal Structures*, Vol. 3, p. 455, Interscience, New York (1966).

The properties of the isopolyniobates(V) and tantalates(V) have recently been reviewed[192]. Physical studies indicate that the polymeric species in aqueous alkaline solution (pH $\simeq$ 11 → 13·5) are $(Nb_6O_{19})^{8-}_{aq.}$ and $(Ta_6O_{19})^{8-}_{aq.}$. Protonation of the niobium species (but not of $(Ta_6O_{19})^{8-}_{aq.}$ occurs at lower values of pH, yielding the species $H(Nb_6O_{19})^{7-}_{aq.}$. At about pH = 4·5, $H_6(N_6O_{18})^{2-}_{aq.}$ is claimed to be formed and to be converted irreversible to niobium pentoxide which precipitates. Recent Raman studies suggest[193] that in 12 M KOH the hexameric species, and not $(NbO_4)^{3-}$, is still predominant.

Niobium forms both normal "8:6" and acid "7:6" salts of the types $M^I_8[Nb_6O_{19}]xH_2O$ (where, for example, M^I = Na or K; $x = 16$) and $M^I_7H[Nb_6O_{19}]xH_2O$ (M^I = Na with $x = 16$ or K with $x = 13$), respectively, but tantalum appears to form only compounds of the former type. The normal salts are obtained by fusion of the pentoxides with potassium hydroxide or carbonate followed by dissolution of the melt and crystallization. Products containing excess alkali have been reported to possess the same X-ray powder patterns and it is suggested that a limited amount of water of crystallization may be replaced by base. Conversion to the acid hexaniobates is achieved by crystallization from water on the addition of alcohol or following vacuum evaporation of the solvent. Mixed hexaniobates(V) and hexatantalates(V) such as $Na_4K_4[Nb_6O_{19}].9H_2O$ and $Na_3K_5[Ta_6O_{19}].10H_2O$ are also known.

Structural studies have shown that the solid compounds contain octahedral groups of metal atoms (metal–metal distance = 3·3 Å) and a symmetrical $[M_6O_{19}]^{8-}$ structure similar to that of the decavanadate ion is proposed. Raman studies indicate that the solid compounds contain the same species as those found in alkaline solutions of the hexaniobates(V) and hexatantalates(V).

In view of the evidence concerning the protonation of $[Nb_6O_{19}]^{8-}_{aq.}$ it is suggested that the soluble metaniobates (1:1 hydrated salts) should be more correctly formulated as salts of the $[H_2(Nb_6O_{19})]^{6-}$i on. It is perhaps significant in this connection that analogous soluble metatantalates are unknown. The soluble sodium orthoniobate (Na_3NbO_4) is irreversibly hydrolysed to the hexaniobate in aqueous solution.

Heteropolyniobates(V) containing manganese(IV) and nickel(IV) have recently been prepared in aqueous solution[194–196]. (These include $Na_{12}[Mn(Nb_{12}O_{38})].50H_2O$ and $K_8Na_4[Ni(Nb_{12}O_{38})].21H_2O$). The mixed compound $Na_{12}[Mn(Ta, Nb)_{12}O_{38}].50H_2O$ (Ta:Nb = 1:1·36) is known but heteropolytantalates(V) have not yet been prepared. In $Na_{12}[Mn(Nb_{12}O_{38})].50H_2O$ the heteropolyanion consists[196] of two isopoly $[Nb_6O_{19}]^{8-}$ units bonded to a central Mn ion which occupies an octahedral pocket in one of their faces. The mean Mn–O distance is 1·87 Å with Nb–O bond lengths ranging from 1·75 to 2·43 Å. Spectral evidence indicates that sodium 6-niobethylene-diamine cobalt(III) hydrate, $Na_5Co(C_5H_8N_2)Nb_6O_{19}.18H_2O$, which is prepared in alkaline solution, contains discrete $Nb_6O^{8-}_{19}$ groups[197].

192 M. T. Pope and B. W. Dale, *Quart. Rev.* **22** (1968) 527.
193 W. P. Griffiths and I. D. Wickins, *J. Chem. Soc.* (A) (1966) 1087.
194 B. W. Dale, J. M. Buckley and M. T. Pope, *J. Chem. Soc.* (A) (1969) 301.
195 C. M. Flynn (Jr.) and G. D. Stucky, *Inorg. Chem.* **8** (1969) 332.
196 C. M. Flynn (Jr.) and G. D. Stucky, *Inorg. Chem.* **8** (1969) 335.
197 C. M. Flynn (Jr.) and G. D. Stucky, *Inorg. Chem.* **8** (1969) 178.

2.8. SULPHIDES, SELENIDES AND TELLURIDES

Niobium and tantalum form a wide range of compounds with the first four elements of Group VI; those formed with oxygen have been discussed in section 2.6. There is no information available on niobium and tantalum polonides. The earlier work on the sulphides, selenides and tellurides is discussed by Fairbrother[73] and for critical reviews on the sulphides (up to 1963) and on the selenides and tellurides (up to 1965) the excellent articles by Jellinek[198] and Selte *et al.*[199] are recommended. References to more recent investigations are given in Tables 23 and 24, respectively. The reliability of different results reported by various laboratories is assessed in the review articles and, in addition, the wide range of nomenclatures used is standardized. Even so these reviewers employ different nomenclatures. Thus, Jellinek[198] designates the various disulphide phases 1*s*-TaS_2, 2*s*-TaS_2, 3*s*-TaS_2, 3*s*-$Ta_{1+x}S_2$, etc., whereas Selte *et al.*[199] prefer the slightly more informative nomenclature 1T-TaS_2, 2H-TaS_2, 3R-$Ta_{1+x}S_2$, etc. The latter has been adopted for the present discussion. The phases designated in this manner all possess trigonal (T), hexagonal (H) or rhombohedral (R) layer structures XMX (X = S, Se or Te; M = Nb or Ta) and the integer corresponds to half the number of X layers, or, alternatively, to the number of complete XMX triple layers (or slabs (s)) per unit length of the c_0-axis. Polymorphs with the same symmetry and values of the integer are designated by the subscripts (a), (b) or (c).

The sulphides, selenides and tellurides are all generally prepared by direct union of the appropriate elements under varying conditions, although oxygen-free disulphides have been obtained by reactions involving the pentoxides and carbon disulphide vapour at 900–1300°C.

Sulphides

The presently known sulphide phases are listed in Table 23 together with available crystallographic data and information concerning the ranges of homogeneity of the different polymorphs of the $M_{1+x}S_2$ phases. These, in general, are derived by the random addition of metal atoms between the MS_2 layers and are very often found to coexist in a given preparation. The remaining phases all appear to have a small range of homogeneity. Whereas the coordination around the metal atom in all the niobium disulphide phases and in the $2H_{(a)}$- and 3R-tantalum disulphide phases is trigonal prismatic, it is reported to be octahedral in 1T-TaS_2 and there are, apparently, alternate slabs with trigonal prismatic and octahedral coordination around the metal atoms in 6R-TaS_2. Thus, the behaviour observed for tantalum is intermediate between that of Ti, Zr and Hf (octahedral coordination) and Nb, Mo and W (trigonal prismatic coordination) disulphides. A similar situation exists for the corresponding polymorphs of tantalum diselenide. Structural information on these and the remaining sulphides is discussed in the references cited in Table 23.

Some unpublished information concerning two apparently new NbS_{1-x} phases is referred to in the recent paper dealing with the preparation and structure of Nb_3S_4 [200], which compound exhibits temperature-independent paramagnetism.

Information on ternary niobium and tantalum sulphide phases is available in the compilation by Pearson[201] (up to 1966) and it was reported more recently that paramagnetic

198 F. Jellinek, *Arkiv Chem.* **20** (1963) 447.
199 K. Selte, E. Bjerkelund and A. Kjekshus, *J. Less-Common Metals,* **11** (1966) 14.
200 A. F. J. Ruysink, F. Kodjik, A. J. Wayner and F. Jellinek, *Acta Cryst.* **24B** (1968) 1614.
201 W. B. Pearson, *Handbook of Lattice Spacings and Structures of Metals,* Vol. 2, Pergamon Press, Oxford (1967).

TABLE 23. CRYSTALLOGRAPHIC DATA FOR THE NIOBIUM AND TANTALUM SULPHIDE PHASES

Compound¶	Range of homogeneity	Symmetry	Space group	Lattice parameters				Reference
				a_0 (Å)	b_0 (Å)	c_0 (Å)	$\beta°$	
$Nb_{2-y}S_2$	$0 \leqslant y \leqslant 0{\cdot}3$	Hexagonal	—	3·32	—	6·46†	—	a
*2N-$Nb_{1+x}S_2$	$0{\cdot}3 \leqslant x \leqslant 0{\cdot}43$	Hexagonal	D_{6h}^4–$P6_3/mmc$	3·31	—	12·53‡	—	a
*3R-$Nb_{1+x}S_2$	$0{\cdot}12 \leqslant x \leqslant 0{\cdot}25$	Rhombohedral	C_{3v}^5–$R3m$	3·33	—	17·80§	—	a
*2H-NbS_2	—	Hexagonal	D_{6h}^4–$P6_3/mmc$	3·31	—	11·89	—	a
*3R-NbS_2	—	Rhombohedral	C_{3v}^5–$R3m$	3·33	—	17·81	—	a
*Nb_3S_4	—	Hexagonal	C_{6h}^2–$P6_3/m$	9·5806	—	3·3747	—	b
*$Nb_{21}S_8$	—	Tetragonal	C_{4h}^5–$I4/m$	16·794	—	3·359	—	c
NbS_3	—	Monoclinic	C_2^2–$P2_1$ or C_{2h}^2–$P2_1/m$	4·94	6·74	18·1	97·5	a
*1T-TaS_2	—	Trigonal	D_{3d}^3–$P\bar{3}ml$	3·36	—	5·90	—	d
*2H$_{(a)}$-TaS_2	—	Hexagonal	D_{6h}^4–$P6_3/mmc$	3·315	—	12·10	—	d
*3R-TaS_2	—	Rhombohedral	C_{3v}^5–$R3m$	3·32	—	17·9	—	d
*6R-TaS_2	—	Rhombohedral	C_{3v}^5–$R3m$	3·335	—	35·85	—	d
TaS (random)	—	Hexagonal	—	1·92	—	5·99	—	d
*2H-$Ta_{1+x}S_2$	$0{\cdot}2 \leqslant x \leqslant 0{\cdot}35$	Hexagonal	D_{6h}^4–$P6_3/mmc$	3·295	—	12·45 ‖	—	d
*3R-$Ta_{1+y}S_2$	$y \approx 0{\cdot}15$	Rhombohedral	C_{3v}^5–$R3m$	3·31	—	18·20	—	d
*6R-$Ta_{1+z}S_2$	$z \approx 0{\cdot}2$	Rhombohedral	—	3·315	—	36·20	—	d
TaS_3	—	Orthorhombic	(?) D_2^5–$C222_1$	36·804	15·173	3·340	—	e
*Ta_2S	—	Orthorhombic	D_{2h}^{11}–$Pbcm$	7·381	5·574	15·195	—	f
*Ta_6S	—	Monoclinic	C_{2h}^6–$C2/c$	14·158	5·284	14·789	118·1	f

* Structural data (sometimes only approximate) are given in the references cited.
¶ For discussion of nomenclature (1T-, 2H-, etc.) see page 600.
† For $y \approx 0{\cdot}3$.
‡ For $Nb_{1\cdot 3}S_2$.
§ For $Nb_{1\cdot 12}S_2$.
‖ For $Ta_{1\cdot 2}S_2$.

[a] F. Jellinek, G. Brauer and H. Müller, *Nature*, **185** (1960) 376.
[b] A. F. J. Ruysink, F. Kodjik, A. J. Wagner and F. Jellinek, *Acta Cryst*. **24B** (1968) 1614.
[c] H. F. Franzen, T. A. Beineke and B. R. Conrad, *Acta Cryst*. **24B** (1968) 412.
[d] F. Jellinek, *J. Less-Common Metals*, **4** (1962) 9.
[e] E. Bjerkelund and A. Kjekshus, *Z. anorg. und allgem. Chem*. **328** (1964) 235.
[f] H. F. Franzen and J. G. Smeggil, *J. Am. Chem. Soc*. **91** (1969) 2814.

TABLE 24. CRYSTALLOGRAPHIC DATA FOR THE NIOBIUM AND TANTALUM SELENIDES AND TELLURIDES

Compound§	Range of homogeneity†	Symmetry	Space group/Structure type	Lattice parameters				Reference
				a_0 (Å)	b_0 (Å)	c_0 (Å)	β°	
*Nb_5Se_4	—	Tetragonal	C^5_{4h}–$I4/m$; Ti_5Te_4	9·871	—	3·4529	—	a
*Nb_3Se_4	—	Hexagonal	C^6_{2h}–$P6_3/m$; Nb_3Te_4	10·012	—	3·4707	—	b
*Nb_2Se_3	—	Monoclinic	C^2_{3h}–$P2_1/m$; Mo_2Se_3	6·503	3·434	9·215	103·39	c
*2H-$NbSe_2$	$0·00 \leqslant x \leqslant 0·29$	Hexagonal	D^4_{6h}–$P6_3/mmc$; NbS_2	3·446	—	12·55	—	d, e
*3R-$NbSe_2$	$x \approx 0·11$	Rhombohedral	C^5_{3v}–$R3m$; —	3·45	—	18·88	—	d
*4H-$NbSe_2$	$0·05 \leqslant x \leqslant 0·17$	Hexagonal	D^4_{6h}–$P6_3/mmc$; NbS_2	3·44	—	25·24	—	d
*$2H_{(b)}Nb_{1+x}Se_2$	$0·05 \leqslant x \leqslant 0·10$	Hexagonal	D^1_{3h}–$P\bar{6}m2$; —	3·454	—	12·58	—	f
α-$NbSe_4$	—	Tetragonal	— ; —	28·75	—	3·836	—	g
β-$NbSe_4$	—	Tetragonal	— ; —	9·503	—	19·15	—	g
1T-$TaSe_2$	—	Trigonal	D^3_{3d}–$P\bar{3}ml$; $Cd(OH)_2$	3·4769	—	6·2722	—	h
*2H-$TaSe_2$	$0·00 \leqslant x \leqslant 0·03$ $0·25 \leqslant x \leqslant 0·64$	Hexagonal	D^4_{6h}–$P6_3/mmc$; NbS_2	3·43	—	12·71	—	d
*$2H_{(b)}$-$Ta_{1+x}Se_2$	$x \approx 0·1$	Hexagonal	D^1_{3h}–$P\bar{6}m2$; —	3·442	—	12·62	—	f
*3R–$TaSe_2$	$0·03 \leqslant x \leqslant 0·25$	Rhombohedral	D^5_{3d}–$R\bar{3}m$; $CdCl_2$	3·428	—	19·100	—	i, j
*$4H_{(a)}$-$TaSe_2$	—	Hexagonal	D^1_{3d}–$P\bar{6}m2$; —	3·43	—	25·55	—	j
*$4H_{(b)}$-$TaSe_2$	—	Hexagonal	D^4_{6h}–$P6_3/mmc$; NbS_2	3·46	—	25·18	—	d
*$4H_{(c)}$-$TaSe_2$	—	Hexagonal	C^4_{6v}–$P6_3/mc$; —	3·436	—	25·532	—	k
6R-$TaSe_2$	$0·03 \leqslant x \leqslant 0·25$	Rhombohedral	C^5_{3v}–$R3m$; TaS_2	3·46	—	37·90	—	j
*$TaSe_3$	—	Monoclinic	C^2_2–$P2_1$ or C^3_{3h}–$P2_1/m$	10·411	3·494	9·836	106·36	l
*Ta_2Se_3	—	Monoclinic	C^2_{3h}–$P2/_1m$; Mo_2S_3	6·495	3·408	9·206	103·63	c
*Nb_5Te_4	—	Tetragonal	C^5_{4h}–$I4/m$; Ti_5Te_4	10·231	—	3·719	—	a
Nb_3Te_4	—	Hexagonal	C^6_{2h}–$P6_3/m$; —	10·671	—	3·6468	—	b
*β-$NbTe_2$	—	Monoclinic	C^3_{2h}–$C2/m$; —	19·39	3·642	9·375	134·59	m
*$NbTe_4$	—	Tetragonal	D^2_{2h}–$P4/mmc$ or C^5_{4v}–$P4cc$	12·998	—	20·511	—	n
$NbTe_4$‡	—	Orthorhombic	— ; —	3·611	13·265	8·681	—	o
*β-$TaTe_2$	$TaTe_2 \rightarrow TaTe$	Monoclinic	C^3_{2h}–C_2/m; —	19·31	3·651	9·377	134·22	m
*$TaTe_4$	—	Tetragonal	D^2_{2h}–$B4/mmc$ or C^5_{4v}–$P4cc$	13·028	—	20·427	—	n, p
$TaTe_4$‡	—	Orthorhombic	— ; —	3·618	13·272	8·685	—	o

* Structural data are provided in the references cited.

§ For discussion of the nomenclature (1T-; 2H-, etc.) see page 600.

† According to recent work by K. Selte, E. Bjerkelund and A. Kjekshus, *J. Less-Common Metals*, **11** (1966) 14.

‡ High pressure modification.

[a] K. Selte and A. Kjekshus, *Acta Chem. Scand.* **17** (1963) 2560.

[b] K. Selte and A. Kjekshus, *Acta Cryst.* **17** (1964) 1568.

[c] F. Kadjik, R. Huisman and F. Jellinek, *Acta Cryst.* **24B** (1968) 1102.

[d] B. E. Brown and D. J. Beernsten, *Acta Cryst.* **18** (1965) 31.

[e] K. Selte and A. Kjekshus, *Acta Chem. Scand.* **18** (1964) 697.

[f] R. Huisman, F. Kadjik and F. Jellinek, *J. Less-Common Metals*, **12** (1967) 423.

[g] K. Selte and A. Kjekshus, *Acta Chem. Scand.* **19** (1965) 1002.

[h] E. Bjerkelund and A. Kjekshus, *Acta Chem. Scand.* **21** (1967) 513.

[i] L. H. Brixner, *J. Inorg. Nuclear Chem.* **24** (1962) 257.

[j] F. Kadjik, R. Huisman and F. Jellinek, *Rev. Trav. Chim.* **83** (1964) 768.

[k] R. Huisman and F. Jellinek, *J. Less-Common Metals*, **17** (1969) 111.

[l] E. Bjerkelund and A. Kjekshus, *Acta Chem. Scand.* **19** (1965) 701.

[m] B. E. Brown, *Acta Cryst.* **20** (1966) 264.

[n] K. Selte and A. Kjekshus, *Acta Chem. Scand.* **18** (1964) 690.

[o] E. Bjerkelund and A. Kjekshus, *Acta Chem. Scand.* **22** (1968) 3336.

[p] E. Bjerkelund and A. Kjekshus, *J. Less-Common Metals*, **7** (1964) 231.

phases of the types $M^{II}Nb_3S_6$ and $M^{II}Ta_3S_6$ (M^{II} = Mn, Fe, Co and Ni) possess a superstructure derived, for example, from the 2H-NbS_2 type, the M^{II} atoms having an ordered distribution over the octahedral holes between the trigonal prismatic NbS_2 layers[202].

Selenides

Direct combination of niobium or tantalum with selenium has led to the characterization of numerous phases (Table 24). As is the case with the disulphides, phases of composition $M_{1+x}Se_2$ are known, the overall range of homogeneity being $0{\cdot}00 \leqslant x \leqslant 0{\cdot}29$ for niobium and $0{\cdot}00 \leqslant x \leqslant 0{\cdot}69$ for tantalum. Individual preparations often contain two or more polymorphs[199]. Apart from those in the 1T- (octahedral) and $4H_{(b)}$- and 6R-$TaSe_2$ phases (both alternate octahedral and trigonal prismatic) the coordination polyhedron around the metal atom in the niobium and tantalum diselenides is trigonal prismatic (cf. the disulphides, page 600). The polymorphism of $TaSe_2$ is discussed in detail by Huisman and Jellinek[203] and structural information on these and other selenides will be found in the references cited (Table 24). Although the existence of three crystal modifications of $NbSe_4$ has been reported, these are incompletely characterized and more information would be of value[199].

$Nb_{1{\cdot}00}Se_2$ is paramagnetic, the observed susceptibility almost satisfying the Curie–Weiss Law above room temperature. However, at about 200°K the material becomes antiferromagnetic. The addition of niobium atoms ($Nb_{1+x}Se_2$) changes the magnetic properties until almost temperature-independent paramagnetism is exhibited[204].

Tellurides

Relatively few tellurides have been characterized (Table 24). $NbTe_2$, in contrast to the corresponding disulphides and diselenides discussed above, has only a narrow range of homogeneity, but it is reported that $TaTe_2$ exhibits a range extending from this composition through to TaTe[205]. Other results, however, are in disagreement with this observation[199]. The discrepancies in the various reports concerning niobium and tantalum telluride phases are discussed in detail by Selte *et al.*[199] and need not be dealt with here.

$NbTe_4$ and $TaTe_4$ are both diamagnetic whilst Nb_3Te_4 and Ta_5Te_4 exhibit weak, almost temperature-independent paramagnetism.

2.9. NITRIDES, PHOSPHIDES, ARSENIDES AND ANTIMONIDES

A wide range of crystalline phases are formed with the elements of Group V. For a critical review of the literature on nitrides the reader is referred to the article by Storms[206] who also discusses the phase diagrams and other transition metal nitrides. Information on the compounds formed with each of the Group V elements is discussed in other articles[73, 201] and summaries are available for the nitrides[207, 208] and arsenides[209]. In addition to the compounds discussed here, carbonitrides, oxonitrides and other ternary phases have been

202 J. M. Van den Berg and P. Cossee, *Inorg. Chim. Acta*, **2** (1968) 143.
203 R. Huisman and F. Jellinek, *J. Less-Common Metals*, **17** (1969) 111.
204 K. Selte and A. Kjekshus, *Acta Chem. Scand.* **19** (1965) 258.
205 B. E. Brown, *Acta Cryst.* **20** (1966) 264.
206 E. K. Storms, U.S. Report LA-2942 (1964).
207 A. Taylor and N. J. Doyle, *J. Less-Common Metals*, **13** (1967) 399.
208 K. Östhagen and P. Kofstad, *J. Less-Common Metals*, **5** (1963) 7.
209 S. Rundqvist, *Nature*, **211** (1966) 847.

studied[73, 201]. Despite the increased metallic character of antimony relative to N, P and As certain of the compounds formed with niobium and tantalum appear to possess non-metallic character and, therefore, are discussed in this section. Only one bismuthide is known, Nb_3Bi and this possesses the β-W type of structure at high pressure[73].

Nitrides

The interstitial adsorption of nitrogen by metallic niobium and tantalum has been extensively studied[73, 201, 206–208]. Terminal solubilities for niobium and tantalum, respectively, are expressed by the equations,

$$C\ (\text{at.}\%\text{N}) = 720 \times \exp\left[\frac{-20{,}000}{T}\right]$$

and

$$C\ (\text{at.}\%\text{N}) = 35 \times \exp\left[\frac{-5{,}400}{RT}\right]$$

whilst the equation for the niobium-nitrogen equilibrium is found to be;

$$C\ (\text{at.}\%\text{N}) = 6{\cdot}2 \times 10^{-4} (\rho_{N2})^{\frac{1}{2}} \exp\left[\frac{46{,}000}{RT}\right].$$

However, below 1000°C the observed solubility for niobium becomes progressively higher than predicted and, for example, is 0·006 wt. % N_2 at 300°C. The niobium–nitrogen system is an excellent example of an ideal interstitial solid solution of nitrogen in a metal lattice[73]. The solid solution phases formed in both the niobium and tantalum systems possess the expanded cubic metal lattice, the lattice parameter (a_0) varying with composition[201,206].

A wide range of crystalline nitride phases, generally obtained by reactions involving the appropriate metal or hydride and nitrogen or gaseous ammonia, are known and available crystallographic data are collected in Table 25 together with information on the ranges of homogeneity of the phases. The variation of the lattice parameters of δ-NbN and β-NbN with composition has been determined[201, 206]. A cubic form of TaN, possibly equivalent to δ′-NbN has been observed[206] in samples heated (1400°C) at high pressure ($\simeq$ 45 kilobar) and there is evidence for the existence of the phase $TaN_{0{\cdot}05}$ [73, 201, 208].

The cubic modifications of the mononitrides are yellow (cf. other cubic nitrides) whereas ε-NbN and ε-TaN are respectively black and blue-grey. Both β-Nb_2N and δ-NbN are insoluble in boiling hydrochloric, nitric or sulphuric acid. The mononitride is oxidized at high temperature in the atmosphere, forming Nb_2O_5, and, when heated, dissolves in strongly alkaline solutions with the evolution of ammonia. β-Nb_2N on the other hand evolves nitrogen under the latter conditions. Few chemical properties have been reported for the tantalum nitrides, but Ta_3N_5, which is dark red in colour, is known to decompose to the mononitride at 500°C in a vacuum.

δ-NbN becomes superconducting at about 16°K and, as a consequence of the relatively high value of this transition temperature (cf. cubic VN, 8·3°K), this property has been extensively studied[206]. The tetragonal form, γ-NbN, is also superconducting, but only at a lower temperature, whereas ε-NbN and the hexagonal forms of TaN are normal above 1·2°K and 1·94°K, respectively, and neither Nb_2N nor Ta_2N are superconducting down to 1·2°K. Selected thermodynamic data are listed in Table 21 and the reader is referred to the article by Storms[206] for details of other physical properties.

Phosphides

Until recently relatively few niobium and tantalum phosphides had been characterized[73], but there is now a total of eight compounds known (Table 25). Monophosphides are

TABLE 25. CRYSTALLOGRAPHIC DATA FOR NIOBIUM AND TANTALUM NITRIDES, PHOSPHIDES, ARSENIDES, ANTIMONIDES AND BISMUTHIDES

Compound	Symmetry	Space group/Structure type	Lattice parameters				Reference
			a_0 (Å)	b_0 (Å)	c_0 (Å)	$\beta°$	
β-Nb_2N§§	Hexagonal	D_{6h}^4–$P6_3/mmc$; W_2C	3·056	—	4·996	—	a
γ-NbN†	Tetragonal	— ; —	4·386	—	4·335	—	b
NbN‡	Hexagonal	D_{3h}^1–$P\bar{6}m2$; WC	2·950	—	2·772	—	b
δ-NbN_{1-x}§	Cubic	O_h^5–$Fm3m$; NaCl	4·388	—	—	—	a
δ'-NbN‖	Hexagonal	D_{6h}^4–$P6_3/mmc$; NiAs	2·968	—	5·549	—	a
ε-NbN¶	Hexagonal	10% $P\bar{6}c2$; 90% $P6_3/mmc$	2·9591	—	11·2714	—	a, b
Ta_2N**	Hexagonal	D_{6h}^4–$P6_3/mmc$; W_2C	3·0420	—	4·9230	—	b
δ-TaN††	Hexagonal	D_{3h}^1–$P\bar{6}m2$; WC	2·925	—	2·876	—	b
ε-TaN‡‡	Hexagonal	D_{6h}^1–$P6/mmm$; CoSn	5·1913	—	2·9108	—	b
Ta_3N_5	Tetragonal	— ; —	10·265	—	3·898	—	c
*Nb_3P	Tetragonal	C_{2h}^4–$P4_2/n$; Ti_3P	10·128	—	5·089	—	d, e
*β-NbP	Tetragonal	C_{4v}^{11}–$I4_1md$; NbAs	3·334	—	11·378	—	e, f
NbP_2	Monoclinic	C_{2h}^3–$C2/m$; $NbAs_2$	8·8715	3·2663	7·5199	119·1	e
*Nb_7P_4	Monoclinic	C_{2h}^3–$C2/m$; —	14·950	3·440	13·848	104·7	e, g
Ta_3P	Tetragonal	C_{2h}^4–$P4_2/n$; Ti_3P	10·154	—	5·012	—	e
β-TaP	Tetragonal	C_{4v}^{11}–$I4_1md$; NbAs	3·319	—	11·341	—	e
TaP_2	Monoclinic	C_{2h}^3–$C2/m$; $NbAs_2$	8·8608	3·2677	7·4882	119·3	e
Ta_5P_4	Orthorhombic	— ; —	14·420	11·547	3·400	—	e
*NbAs	Tetragonal	C_{4v}^{11}–$I4_1md$	3·4517	—	11·680	—	f
*$NbAs_2$	Monoclinic	C_{2h}^3–$C2/m$;	9·357	3·3823	7·792	119·46	h
Nb_3As	Tetragonal	C_{2h}^4–$P4_2/n$; Ti_3P	10·289	—	5·191	—	i
*TaAs	Tetragonal	C_{4v}^{11}–$I4_1md$; NbAs	3·437	—	11·656	—	f
*$TaAs_2$	Monoclinic	C_{2h}^3–$C2/m$; $NbAs_2$	9·350	3·388	7·760	119·75	j
Ta_3As	Tetragonal	C_{2h}^4–$P4_2/n$; Ti_3P	10·301	—	5·110	—	i
Nb_3Sb	Cubic	O_h^3–$Pm3m$; β-W	5·2643	—	—	—	a
Nb_5Sb_4	Tetragonal	C_h^5–$I4/m$; Ti_5Te_4	10·314	—	3·5566	—	a
*$NbSb_2$	Monoclinic	C_{2h}^3–$C2/m$; $NbAs_2$	10·239	3·6319	8·333	120·07	a
Ta_3Sb	Cubic	O_h^3–$Pm3m$; β-W	5·2595	—	—	—	a
Ta_5Sb_4	Tetragonal	C_h^5–$I4/m$; Ti_5Te_4	10·248	—	3·5460	—	a
$TaSb_2$	Monoclinic	C_{2h}^3–$C2/m$; $NbAs_2$	10·2218	3·6447	8·2915	120·39	a
Nb_3Bi	Cubic	O_h^3–$Pm3m$; β-W	5·320	—	—	—	a

* Structural data are given in references cited.
§§ Range of homogeneity, $NbN_{0.40}$–$NbN_{0.50}$; parameters quoted for $NbN_{0.50}$.
† Range of homogeneity, $NbN_{0.75}$–$NbN_{0.79}$; parameters quoted for $NbN_{0.79}$.
‡ Range of homogeneity, $NbN_{0.80}$–$NbN_{0.90}$; parameters quoted for $NbN_{0.80}$; may be stabilized by impurities such as oxygen.
§ Range of homogeneity, $NbN_{0.38}$–$NbN_{0.98}$; parameters quoted for $NbN_{0.92}$.
|| Metastable phase; parameters quoted for $NbN_{0.98}$.
¶ Range of homogeneity $NbN_{1.00}$–$NbN_{1.018}$; parameters quoted for $NbN_{1.00}$.
** Approximate range of homogeneity $TaN_{0.36}$–$TaN_{0.5}$; parameters quoted for $TaN_{0.5}$.
†† Range of homogeneity $TaN_{0.80}$–$TaN_{0.90}$; parameters quoted for $TaN_{0.80}$; unconfirmed phase.
‡‡ Range of homogeneity $TaN_{0.98}$–$TaN_{1.00}$; parameters quoted for $TaN_{1.00}$.

[a] W. B. Pearson, *Handbook of Lattice Spacings and Structures of Metals*, Vol. 2, Pergamon Press, Oxford (1967).
[b] E. K. Storms, U.S. Report LA-2942 (1964).
[c] G. Brauer and J. R. Weidlein, *Angew. Chemie, Int. Ed.* **4** (1965) 241.
[d] P. C. Nawapang, *Acta Chem. Scand.* **20** (1966) 2737.
[e] S. Rundqvist, *Nature*, **211** (1966) 847.
[f] H. Boller and E. Parthé, *Acta Cryst.* **16** (1963) 1095.
[g] S. Rundqvist, *Acta Chem. Scand.* **20** (1966) 2427.
[h] S. Furuseth and A. Kjekshus, *Acta Cryst.* **18** (1965) 320.
[i] E. Ganglberger, H. Nowotny and F. Benasovsky, *Monatsh. Chem.* **97** (1966) 1696.
[j] S. Furuseth, K. Selte and A. Kjekshus, *Acta Chem. Scand.* **19** (1965) 95.

conveniently prepared by direct union of the elements and it has been stated that such reactions occur more readily in the presence of iodine. The remaining compounds have been obtained by arc melting mixtures of the appropriate metal and monophosphide[209]. It is now accepted that β-NbP (and the tantalum analogue) possesses the NbAs-type of structure and that the phosphorus-deficient phases, for example $NbP_{0.95}$, possess a disordered form of this structure. The homogeneity range of NbP is $NbP_{1.00}$ to $NbP_{0.81}$; the other phases possess only narrow ranges of homogeneity. Crystallographic data are listed in Table 25 where references to structural properties are given. There appear to be no thermodynamic data available for these phosphides. They are greyish-black, relatively inert materials, being unaffected by water or dilute acids at room temperature. When heated in air they are converted to the pentoxides.

Arsenides

Mono- and diarsenides are known for both niobium and tantalum and the phases Nb_3As and Ta_3As have been characterized recently[210]. Details of the earlier work are discussed by Fairbrother[73]. The compounds are prepared by heating together stoichiometric quantities of the appropriate elements in a vacuum at 800–1200°C. Crystallographic data are listed in Table 25 from which it will be seen that the monophosphides, disulphides and diantimonides possess the structure determined for their arsenide analogues. Nb_3As and Ta_3As possess the Ti_3P-type of structure.

NbAs and $NbAs_2$ are metallic conductors of the *p*- and *n*-type, respectively. The mono- and diarsenides are diamagnetic with no unpaired *d*-electrons present on the niobium or tantalum atoms suggesting the presence of covalent bonding. Thermodynamic data are not available for these compounds.

Antimonides

Three intermediate phases have been reported for both the Nb–Sb and Ta–Sb systems[73] (Table 25), but more recent studies have indicated the existence of Nb_3Sb, Nb_3Sb_2, NbSb and Nb_4Sb_5 in the former system[64]. Obviously further work is required to resolve these discrepancies. As may be expected from the increasing metallic character observed with increasing atomic number of the Group V elements certain of the niobium and tantalum antimonides appear to possess metallic-type bonding. Thus, whereas the diantimonides, which are isostructural with the diarsenides (Table 25), are diamagnetic, phases of the types M_3Sb and M_5Sb_4 (M = Nb and Ta) are weakly paramagnetic and the former are not isostructural with their arsenide analogues. Little is known concerning the chemistry of these compounds.

2.10. CARBIDES AND SILICIDES

Niobium and tantalum carbides and silicides have been extensively studied. Storms[206] has critically reviewed the carbide literature available to 1964 and more recent articles have been written by Frad[211] (carbides) and Fairbrother[73] (carbides and silicides). Summaries of the crystallographic, physical and mechanical properties of these compounds are also available[201, 212] and the information on ternary carbide and silicide phases, which will not be discussed here, has been reviewed[201, 211].

[210] E. Ganglberger, H. Nowotny and F. Benesovsky, *Monatsh. Chem.* **97** (1966) 1696.

[211] W. A. Frad, Chapter III in *Advances in Inorganic and Radiochemistry*, Ed. A. G. Sharpe and H. J. Emeléus, Vol. XI, Academic Press, London (1968).

[212] *Refractory Ceramics for Aerospace*, compiled and edited by J. R. Hague, J. F. Lynch, H. Rudnick, F. C. Holden and W. H. Duckworth, Amer. Ceram. Soc. Publication (1964).

Relatively little is known concerning the binary alloys formed with the more metallic members of Group IV (Ge, Sn and Pb). Such phases are discussed by Pearson[201] and in the references cited in section 2.2 and the only comments which will be made here concern the similarities and differences in their structures relative to those of analogous compounds formed with Group V and other Group IV elements. Thus, unlike the analogous silicides, phosphides and arsenides (section 2.9), phases of the type $M_3^VM^{IV}$ (M^V = Nb and Ta; M^{IV} = Ge, Sn and Pb) all possess the β-W type of structure as do, for example, Nb_3Sb, Ta_3Sb and Nb_3Bi (section 2.9). However, there are, apparently, similarities between the structures exhibited by Nb_5Ge_3 and Ta_5Ge_3 and those of the corresponding polymorphic silicides.

Carbides

Three carbide phases have been characterized for both niobium and tantalum. Details of the niobium–carbon phase diagram are discussed by Storms[206] who also presents a tentative tantalum–carbon phase diagram. Diffusion of carbon into the metal lattice results in the formation of solid solutions which retain the body-centred cubic lattice of the metal, a_0 increasing with carbon content. Approximate solubility limits, which decrease with decreasing temperature, are 2 at. % at 2000°C for Nb and 6 at. % at 2700°C for Ta. Carbides of the types M_2C and MC (M = Nb and Ta) are obtained by a variety of methods including direct union of the components at high temperature, reduction of the oxides by carbon or carbon–hydrogen mixtures, reduction of the pentahalide on a hot carbon filament and by heating the elements in a hydrocarbon–hydrogen mixture. NbC, but not TaC, may be deposited directly from the gas phase onto a hot surface.

Crystallographic data for the M_2C and MC phases are given in Table 26. Nb_2C has a very narrow range of homogeneity at room temperature, but the range of Ta_2C is not very well established[206]. NbC and TaC are both homogeneous phases from approximately $MC_{0.70}$ to $MC_{0.99}$. The lattice parameters for their cubic unit cells vary with carbon content and are represented by the equations

$$a_0(\text{NbC}) = 4{\cdot}09847+0{\cdot}71820(\text{C/Nb})-0{\cdot}34570(\text{C/Nb})^2$$

and

$$a_0(\text{TaC}) = 4{\cdot}3007+0{\cdot}1563(\text{C/Ta})$$

where C/Nb and C/Ta are mole ratios. Other MC phases designated δ-NbC, ε-NbC and ε-TaC have been observed on occasions but they are not well characterized[206]. The presence of traces of oxygen and nitrogen was recently reported to stabilize orthorhombic modifications of niobium subcarbide[213] (β′-Nb_2C).

$TaC_{0.99}$ and $NbC_{0.99}$, which are respectively yellow and lavender-grey in colour, are high melting, relatively inert materials which are unaffected by aqua-regia. They do, however, dissolve in hydrofluoric acid–nitric acid mixtures and it is reported that when heated in a nitrogen–hydrogen mixture the niobium compound is converted to the nitride. Both solids burn in oxygen at moderate temperatures.

The temperature at which NbC and TaC become superconducting varies with the composition of the phase. For example, $NbC_{0.977}$ and $NbC_{0.829}$ become superconducting at 11·1°K and 1·05°K, respectively, whilst the temperatures for $TaC_{0.987}$ and $TaC_{0.848}$ are respectively 9·7°K and 2·04°K[206]. Neither Nb_2C nor Ta_2C are superconducting down to

[213] S. I. Alyamovskii, G. P. Shveikin, P. V. Gel'd and N. M. Volkova, *Russ. J. Inorg. Chem.* **12** (1967) 301.

2·0°K. Selected thermodynamic properties are listed in Table 21; details of other physical properties, including discussions of the conflicting information on vaporization of the monocarbides, are presented in earlier reviews[73, 206, 211, 212].

Because of their high melting points, hardness and chemical inertness the niobium and tantalum carbides are potentially useful materials in the fields of high-temperature technology.

TABLE 26. CRYSTALLOGRAPHIC DATA FOR NIOBIUM AND TANTALUM CARBIDES AND SILICIDES

Compound	Symmetry	Space group/ Structure type	Lattice parameters (Å) a_0	b_0	c_0	Reference
α-Nb_2C	Orthorhombic	Similar to γ-Fe_2M	12·36	10·89	4·96	a
*β-Nb_2C†	Hexagonal	— ; ε-Fe_2N	5·407	—	4·960	b
γ-Nb_2C	Hexagonal	D^4_{6h}–$P6_3/mmc$; W_2C	3·117	—	4·962	a, b
*Nb_2C	Orthorhombic	C^9_{2v}–Pna	10·92	4·974	3·090	c
NbC_{1-x}‡	Cubic	O^5_h–$Fm3m$; NaCl	4·4691	(at $NbC_{0.92}$)	—	b
*β-Ta_2C§	Hexagonal	D^3_{3a}–$P\bar{3}ml$;	3·103	—	4·938	d
TaC_{1-x}‡	Cubic	O^5_h–$Fm3m$; NaCl	4·4555	(at $TaC_{0.994}$)	—	b
Nb_4Si	Hexagonal	— ; —	3·59	—	4·46	b
*Nb_3Si	Cubic	O^1_h–$Pm3m$; $AuCu_3$	4·211	—	—	e
Nb_3Si	Tetragonal	C^4_{2h}–$P4_2/n$; Ti_3P	10·230	—	5·189	c
*β-Nb_5Si_3	Tetragonal	D^{18}_{4h}–$I4/mcm$; W_5Si_3	10·026	—	5·0717	b
*α-Nb_5Si_3	Tetragonal	D^{18}_{4h}–$I4/mcm$; Cr_5Bi_3	6·570	—	11·884	b
$NbSi_2$	Hexagonal	D^1_6–$P6_222$; $CrSi_2$	4·797	—	6·592	b
Ta_2Si	Tetragonal	D^{18}_{4h}–I_4/mcm; Al_2Cu	6·157	—	5·039	b
*β-Ta_5Si_3	Tetragonal	D^{18}_{4h}–I_4/mcm; W_5Si_3	9·88	—	5·06	b
*α-Ta_5Si_3	Tetragonal	D^{18}_{4h}–I_4/mcm; Cr_5Bi_3	6·516	—	11·873	b
$TaSi_2$	Hexagonal	D^1_6–$P6_222$; $CrSi_2$	4·7821	—	6·5695	b

* Structural data are given in references cited.
† Cell refers to a super-structure.
‡ Lattice parameter varies with composition.
§ This is equivalent to the γ-Nb_2C phase in a and b.
[a] E. Rudy and C. E. Brukl, *J. Am. Ceram. Soc.* **50** (1967) 265.
[b] W. B. Pearson, *Handbook of Lattice Spacings and Structures of Metals*, Vol. 1 (1958) and Vol. 2 (1967) Pergamon Press.
[c] H. Nowotny and K. Seifert, *Atomic Energy Rev.*, Special Issue, No. 2, 71 (1968).
[d] A. L. Bowman, T. C. Wallace, J. L. Yarnell, R. G. Wenzel and E. K. Storms, *Acta Cryst.* **19** (1965) 6.
[e] F. Galasso and J. Pyle, *Acta Cryst.* **16** (1963) 228.

Silicides

Both niobium and tantalum react exothermically with silicon and a wide range of silicides have been identified for each element. They may also be deposited from the vapour phase by the reaction between the appropriate pentachloride and silicon tetrachloride in the presence of hydrogen. The phase diagrams have been studied by several groups [60, 62, 73]. In addition to those compounds listed in Table 26 a phase of composition $Ta_{4.5}Si$ is known, and in the presence of traces of carbon, hexagonal modifications of the M_5Si_3 phases are stabilized[201].

Niobium and tantalum disilicides are unaffected by mineral acids, with the exception of hydrofluoric acid, or by dilute alkaline solutions. They are resistant to oxidation in the air up to at least 1000°C.

2.11. BORIDES

Because of the metallic character of the last four Group III elements (Al, Ga, In and Tl) only those compounds formed with boron are discussed here. Phase diagrams and compound formation with aluminium and gallium are dealt with in appropriate reviews[60, 62, 64] Alloys containing indium and thallium appear to be unknown.

For reviews of the literature on niobium and tantalum borides the articles by Fairbrother[73] and Peshev *et al.*[214] are recommended. The most satisfactory preparative technique involves direct union of the elements although reduction of the pentoxides by boron or boron–boron carbide mixtures and electrolysis of molten oxide–calcium fluoride mixtures has been employed. Deposition on a heated surface from the vapour phase mixture of the appropriate pentachloride, hydrogen and boron tribromide results in initial products containing either free niobium or tantalum. Details of the phase diagrams are discussed in recent articles[62, 64, 73].

TABLE 27. CRYSTALLOGRAPHIC DATA FOR NIOBIUM AND TANTALUM BORIDES[(a)]

Compound	Symmetry	Space group/ Structure type	Lattice parameters (Å)		
			a_0	b_0	c_0
Nb_3B_2	Tetragonal	D_{4h}^5–$P4/mbm$; U_3Si_2	3·192	—	3·289
NbB	Orthorhombic	D_{2h}^{17}–$Cmca$; CrB	3·292	8·713	3·165
β-NbB	Cubic (?)	; —	4·210	—	
Nb_3B_4	Orthorhombic	D_{2h}^{15}–$Immm$; Cr_3B_4	3·305	14·08	3·137
NbB_2	Hexagonal	D_{6h}^1–$C6/mmm$; AlB_2	3·110	—	3·268
$Ta_{2\cdot4}B$[(b)]	Tetragonal	D_{4h}^{18}–I_4/mcm; $CuAl_2$	5·7795	—	4·8645
Ta_3B_2	Tetragonal	D_{4h}^5–$P4/mbm$; U_3Si_2	6·176	—	3·2844
TaB	Orthorhombic	D_{2h}^{17}–$Cmca$; CrB	3·276	8·669	3·157
Ta_3B_4	Orthorhombic	D_{2h}^{15}–$Immm$; Cr_3B_4	3·284	13·980	3·129
TaB_2	Hexagonal	D_{6h}^1–$P6/mmm$; AlB_2	3·088	—	3·241

[a] W. B. Pearson. *Handbook of Lattice Spacings and Structures of Metals*, Pergamon Press, Vol. I (1958) and Vol. II (1967).
[b] This phase has the composition Ta_2B according to L. Brewer, D. L. Sawyer, D. H. Templeton and C. H. Dauben, *J. Am. Ceram. Soc.* **34** (1957) 173.

The presently known borides are listed in Table 27 together with available crystallographic data. Structural properties are discussed by Fairbrother[73]. Phases of composition MB and M_3B_4 possess only narrow ranges of homogeneity, but the diborides exhibit quite a wide range. Niobium and tantalum borides, like the carbides and nitrides, are hard, refractory materials which are metallic conductors and, therefore, are of potential value in the fields of high temperature technology. The mono- and diborides melt congruently, but there is evidence that the latter phases undergo decomposition just below their melting points. Some thermodynamic properties are listed in Table 21.

[214] P. Peshev, L. Leyarovska and G. Bliznakov, *J. Less-Common Metals*, **15** (1968) 259.

2.12. NITRATES AND NITRATO-COMPLEXES

The little information available on niobium and tantalum nitrates relates only to pentavalent compounds and there appears to have been little serious effort to prepare tetravalent nitrates or nitrato-complexes.

Nitrates and Oxynitrates

Simple pentanitrates are unknown, the reaction between liquid dinitrogen pentoxide and the pentachlorides resulting in the formation of the moisture-sensitive oxytrinitrates $NbO(NO_3)_3$ and $TaO(NO_3)_3$, respectively, both of which are soluble in an excess of the nitrating agent[215, 216]. Niobium and tantalum pentoxide are unaffected by liquid dinitrogen pentoxide, but the freshly precipitated, vacuum-dried hydroxides do react slowly to yield the oxytrinitrates. The nitrate groups are covalently bonded and for $NbO(NO)_3$ the Nb=O stretching vibration is reported at 906 cm^{-1}. This compound possesses tetragonal symmetry with $a_0 = 11{\cdot}92$ and $c_0 = 5{\cdot}33$ Å.

Liquid dinitrogen tetroxide does not react with niobium or tantalum pentachloride, except in the presence of ionizing solvents, or with metallic niobium and tantalum or their oxides under any conditions. The reaction with niobium pentachloride in anhydrous methyl cyanide results in the formation of $NbO_2NO_3.0{\cdot}67MeCN$ ($Nb_3O_6.NO_3.2MeCN$), which is a white, moisture-sensitive complex, but irreproducible behaviour is reported for tantalum pentachloride[216]. Similar oxynitrate complexes with N,N-dimethylacetamide and nitromethane have been reported, but such compounds are not well characterized.

Nitrato-complexes

$NMe_4NbO(NO_3)_4$ and $NMe_4TaO(NO_3)_4$, the only known nitrato-complexes, are obtained by the addition of liquid dinitrogen pentoxide to the appropriate hexachloro-complex followed by removal of excess reagent by vacuum evaporation at room temperature[217]. Alternatively, $NMe_4NbO(NO_3)_4$ may be prepared by passing gaseous dinitrogen tetroxide through a solution of the hexachloro-complex in methyl cyanide, behaviour which contrasts with that discussed above for reactions involving the pentachloride. Attempts to prepare caesium salts by similar reactions have been unsuccessful. Infrared studies indicate that the nitrate groups are covalently bonded and Nb=O and Ta=O stretching vibrations are reported to occur at 881 and 877 cm^{-1}, respectively. Structural information is not available.

In contrast to the behaviour of niobium(V) and tantalum(V) similar reactions involving hexachloroprotactinates(V), including $CsPaCl_6$, lead to the formation of hexanitrato-complexes[217], presumably as a consequence of the somewhat larger ionic radius of Pa(V), 0·90 Å.

2.13. PHOSPHITES AND PHOSPHATES

Niobium and tantalum phosphites are unknown, but both elements form pentavalent metaphosphates and normal phosphates[73]. The former class of compound, M^VOPO_4 or $M^V_2O_5.P_2O_5$ (M^V = Nb and Ta), is obtained by ignition of the precipitate formed, for example, on the addition of an excess of phosphoric acid–nitric acid solution to potassium

215 B. O. Field and C. J. Hardy, *Proc. Chem. Soc.* (1963) 11.
216 K. W. Bagnall, D. Brown and P. J. Jones, *J. Chem. Soc.* (A) (1964) 2396.
217 D. Brown and P. J. Jones, *J. Chem. Soc.* (A) (1966) 733.

niobate(V) or tantalate(V). A more unusual method, which results in the formation of single crystals of $NbOPO_4$, consists of heating together phosphoric acid and Nb_2O_5 in a gold capsule at 600°C at a pressure of 1800 atm[218]. The compound PNb_9O_{25} has been obtained by a similar technique[219]. $NbOPO_4$ possesses tetragonal symmetry, space group $P4/n$ with $a_0 = 6{\cdot}3873$ and $c_0 = 4{\cdot}1037$ Å and the structure comprises chains of corner-shared NbO_6 octahedra running parallel to c_0 and linked together by PO_4 tetrahedra. The niobium atom is surrounded by six oxygen atoms, four at a distance of 1·969 Å, one at 1·78 Å and a sixth at 2·32 Å. This type of distortion, involving movement of the niobium towards one of the corner-shared oxygen atoms, is similar to that observed for PNb_9O_{25} [219] and Nb_3O_7F[220].

In the presence of controlled amounts of phosphoric acid the hydrated compound $2Nb_2O_5.P_2O_5.6H_2O$, a white, amorphous solid, is precipitated when aqueous solutions containing hydrofluoric acid and phosphoric acid are boiled[221]. Thermal decomposition leads to the formation of $2Nb_2O_5.P_2O_5.4H_2O$ (105°) and then $2Nb_2O_5.P_2O_5$ (1000°C), which is insoluble in water and concentrated nitric acid but dissolves readily in concentrated phosphoric acid. The analogous tantalum compound $2Ta_2O_5.P_2O_5$ is also known[222].

The normal phosphates $Nb_3(PO_4)_5$ and $Ta_3(PO_4)_5$ crystallize when hydrous niobium oxide or tantalum pentachloride, respectively, are heated with an excess of fused H_3PO_4 [223]. Excess phosphoric acid is removed by igniting the initial products at 1200°. Little has been reported concerning the chemistry of these interesting compounds and no structural information is available.

The methylphosphinates $Nb_2O(O_3PCH_3)_4$ and $Ta_2O(O_3PCH_3)_4$ precipitate when methylphosphinic acid and the appropriate pentachloride are heated together in acetone[224]. It is reported that these compounds, which are insoluble in water and organic solvents, contain M–O–M–O linkages (ν_{M-O} at 780–800 cm^{-1}).

2.14. SULPHITES, SULPHATES AND THIOSULPHATES

$Ta_2(SO_4)_5$ is known and several pentavalent oxysulphates have been characterized for both niobium and tantalum, but sulphites and thiosulphates appear to be unknown[73]. Niobium but not tantalum forms a lower valent sulphate. The literature concerning niobium sulphates is summarized in a recent paper[225].

The lower, apparently mixed, valence state sulphate originally reported[73] as $HNb_3(SO_4)_6.nH_2O$ is more accurately represented by the formula $Nb_6O_3(SO_4)_8.12H_2O$[226]. The niobium atoms possess an average oxidation state of 3·67 in this and the related complexes such as $K_8Nb_6O_3(SO_4)_{12}.21H_2O$ and it is reported that the magnetic properties of this latter complex are consistent with the presence of Nb–Nb bonds with two paramagnetic (μ_{eff} = 1·4 B.M. per niobium atom) and four diamagnetic niobium atoms[227]. Reddish-brown crystals of $K_8Nb_6O_3(SO_4)_{12}.21H_2O$ and similar complex sulphates, which are more

[218] J. M. Longs and P. Kierkegaard, *Acta Chem. Scand.* **20** (1966) 72.
[219] R. S. Roth, A. D. Wadsley and S. Andersson, *Acta Cryst.* **18** (1965) 643.
[220] S. Andersson, *Acta Chem. Scand.* **18** (1964) 2339.
[221] Ya. Rakhmatulin and V. P. Gladyshev, *Russ. J. Inorg. Chem.* **13** (1968) 1351.
[222] A. P. Shtin, *Izv. Sibir. Otdel. Nauk SSSR*, **68** (1961), acc. to *C.A.* **56**, 6873b.
[223] G. B. Seifer and I. V. Tananaev, *Russ. J. Inorg. Chem.* **8** (1963) 30.
[224] Ya. Manevich and Yu. A. Busalev, *Inorg. Materials*, **3** (1967) 1312.
[225] J. E. Land and J. R. Sanchez-Caldas, *J. Less-Common Metals*, **13** (1967) 233.
[226] Ya. G. Goroshchenko and M. I. Andreeva, *Russ. J. Inorg. Chem.* **11** (1966) 1197.
[227] E. I. Krylov and N. N. Kalugina, *Russ. J. Inorg. Chem.* **4** (1959) 1139.

stable than the parent acid, are obtained by addition of the appropriate metal sulphate to a sulphuric acid solution of the parent acid prepared by electrolytic reduction of niobium(V) at a lead or mercury cathode. The potassium salt possesses monoclinic symmetry with $a_0 = 10{\cdot}16$, $b_0 = 17{\cdot}61$, $c_0 = 10{\cdot}03$ Å and $\beta = 58{\cdot}93°$.

Niobium(V) and tantalum(V) form somewhat different series of sulphates and oxysulphates and the behaviour of both elements differs from that reported for protactinium(V)[228]. For example[73], the pentachlorides react with sulphur trioxide in sulphuryl chloride solution to form the white, insoluble compounds $Nb_2O(SO_4)_4$ and $Ta_2(SO_4)_5$, respectively, and, whereas the white niobium oxysulphates $Nb_2O(SO_4)_4$, $Nb_2O_2(SO_4)_3$, $Nb_2O_3(SO_4)_2$ and $Nb_2O_4(SO_4)$ can be isolated from the Nb_2O_5–SO_3–H_2O system[229] at SO_3 concentrations ranging from 10–90%, only $Ta_2(SO_4)_5$ and $Ta_2O_4(SO_4).3H_2O$ (together with a phase designated $Ta_2O_5.SO_3.nH_2O$) are obtained from similar solutions containing tantalum(V)[230]. A blue isomer of $Nb_2O_3(SO_4)_2$ prepared, for example, by the addition of the white compound to a boiling aqueous solution of K_2S, has been reported[225] and it is suggested that this may be a pyrosulphate, $Nb_2O_4(S_2O_7)$. It was recently reported that evaporation of sulphuric acid solutions at 300°C yields the hydrated oxysulphates $Ta_2O_3(SO_4)_2.3H_2O$, $Ta_2O_3(SO_4)_2.4H_2O$ or $Ta_2O_3(SO_4)_3.5H_2O$ depending on the treatment of the initial product[231].

Thermal decomposition of $Nb_2O_2(SO_4)_3$ above 540° is stated to result in the successive formation of $Nb_2O_3(SO_4)_2$ and $Nb_2O_4(SO_4)$ although the formation of $Nb_2O_4(SO_4)$ under these conditions has been questioned[225]. $Ta_2(SO_4)_5$, which is appreciably less stable, loses SO_3 above 100°C and is converted to $Ta_2O_3(SO_4)_2$ between 200° and 320°C. This, in turn, loses SO_3 above 340°, decomposing directly to the pentoxide. No structural information is available for any of these pentavalent sulphates and it has been suggested that $Ta_2(SO_4)_5$ and $Nb_2O(SO_4)_4$ are better represented as oxygenated derivatives of polysulphuric acids, viz. $(TaO_2)_2S_5O_{16}$ and $(NbO_2)_2S_4O_{13}$, respectively[232].

Several pentavalent sulphato-complexes, all of which are moisture-sensitive, have been prepared from aqueous sulphuric acid solutions. These include, for example[73], $NH_4NbO(SO_4)_2$, $(NH_4)_3Nb(SO_4)_4$ and $(NH_4)_3Ta(SO_4)_4$.

2.15. SELENITES AND SELENATES

The addition of a twentyfold excess of selenous acid to niobium(V) in hydrochloric acid (1–6 M) or sulphuric acid solution (1–3 M) results in the precipitation of the oxyselenite $Nb_2O_3(SeO_3)_2$ [233]. With smaller amounts of selenous acid or in the presence of ammonium sulphate products of variable composition have been observed to form. No precipitation occurs from hydrofluoric or oxalic acid solution. $Nb_2O_3(SeO_3)_2$ may also be prepared by heating freshly precipitated niobium(V) hydroxide with selenous acid on a water bath. Nothing is known concerning the chemical properties of this compound.

Tantalum(V) selenites are unknown and there appears to have been no attempts to prepare tetravalent selenites or tetra- or pentavalent selenates for either niobium or tantalum.

[228] K. W. Bagnall, D. Brown and P. J. Jones, *J. Chem. Soc.* (A) (1966) 176.
[229] Ya. G. Goroshchenko and M. I. Andreeva, *Russ. J. Inorg. Chem.* **8** (1963) 505.
[230] Ya. G. Goroshchenko and M. I. Andreeva, *Russ. J. Inorg. Chem.* **10** (1965) 517.
[231] R. D. Maslennikova and A. S. Chernyak, *Russ. J. Inorg. Chem.* **13** (1968) 335.
[232] A. D. Woolf, *Chem. and Ind.* (1954) 1320.
[233] T. T. Mityareva, B. I. Danil'tseva and I. A. Sheka, *Russ. J. Inorg. Chem.* **13** (1968) 337.

2.16. COMPLEXES

Halogeno- and oxyhalogeno-complexes are discussed in section 2.4 together with the numerous tetrahalide and pentahalide complexes of the general types $MX_4.2L$ and $MX_5.L$, etc. (L = oxygen, sulphur, selenium and nitrogen donor ligands) whilst nitrato-, sulphato-, thiocyanato- and selenocyanato-complexes are dealt with in sections 2.12, 2.14 and 2.5, respectively. Pentavalent niobium and tantalum form numerous other complexes with, for example, alcohols, hydroxy acids, amines and the chelates such as tropolone, β-diketones and dialkyldithiocarbamates. Relatively few tetravalent, and no trivalent, complexes are known. Examples of complexes with selenium donors appear to be confined to the pentahalide complexes and selenocyanates described earlier (pages 582 and 591, respectively). A notable feature of the chemistry of the complexes described below is the almost total absence of structural information.

Dipyridil and Phthalocyanine Complexes

Tris-2,2′-dipyridilniobium(0), $NbDipy_3$, is obtained by the reduction of niobium pentachloride by dilithium dipyridil in tetrahydrofuran solution. The violet crystals which deposit from the intensely blue solution are paramagnetic, μ_{eff} = 1·75 B.M., and decompose immediately on exposure to air[234].

It is reported[235] that in the presence of quinoline, phthalonitrile reacts with niobium and tantalum pentachloride to yield the monophthalocyanines NbPc and TaPc, behaviour which contrasts with that of protactinium pentaiodide which is converted to the bis-compound $PaPc_2$ [236]. Further information on these compounds would be of value.

Tetravalent

Paramagnetic niobium(IV) tetrakis-β-diketone complexes of the type NbL_4 (L = benzoyltrifluoroacetone, thenoyltrifluoroacetone and dibenzoylmethane) and analogous tropolone and 8-hydroxyquinoline complexes are obtained from the diamagnetic tetrachloride by reactions in non-aqueous solvents[237]. Their effective magnetic moments (1·60–1·66 B.M.) are intermediate between those of the eight-coordinate diarsine complexes, $MX_4.2L$ (X = halide, L = diarsine) and those of the octahedral complexes formed with monodentate ligands. Structural information is not available for these complexes, but on the basis of spectral results it is suggested that the β-diketonates are dodecahedral. The apparently nine-coordinate acetylacetonate complex, $Nbacac_4$.dioxan, possesses the spin only magnetic moment of 1·73 B.M.

It is reported[238] that niobium pentachloride, in contrast to tantalum pentachloride, is reduced in the presence of lithium dialkylamides with the formation of tetravalent complexes such as $Nb(NR_2)_4$ (R = Me, Et, ^nPr, ^nBu, etc.), the extent of reduction increasing with the length of the alkyl chain. Although it is well established that niobium(V) is more easily reduced than tantalum(V) it is believed that steric effects play an important part in this instance since $Nb(NMeBu)_5$ and $Nb(NC_5H_{10})_5$ are both stable (C_5H_{10} = piperidine). $Ta(NEt_2)_4$ and $Ta(NMeBu)_4$ are observed as by-products of the thermal decomposition of the pentakis-complexes but with longer-chain dialkylamides the principle products are of

234 S. Herzog and R. Schuster, *Z. Naturforsch.* **17B** (1962) 62.
235 Yu. A. Buslaev, A. A. Kuznetsova and L. F. Goryachova, *Inorg. Materials*, **3** (1967) 1488.
236 F. Lux, D. Brown, D. Dempf, R. D. Fisher and W. Hagenberg, *Proc. 3rd Int. Pa Conf.*, to be published.
237 R. L. Deutscher and D. L. Kepert, *Chem. Commun.* (1969) 121.
238 D. C. Bradley and R. M. Thomas, *Can. J. Chem.* **40** (1962) 449.

the type $RN{=}Ta(NR_2)_3$ [239]. The niobium(V) pentakisdialkylamides (R = Me and Et) react with carbon disulphide in cyclohexane to yield[240] tetravalent dialkyldithiocarbamates of general formula $Nb(S_2CNR_2)_4$ whereas $Ta(NMe_2)_5$ is not reduced, the product being $Ta(S_2CNMe_2)_5$. Analogous Ti(IV), V(IV) and Zr(IV) dialkyldithiocarbamates are also obtained by similar carbon disulphide insertion reactions and it would be interesting to have structural information on these potentially eight-coordinate sulphur complexes for comparison with that recently reported for the diethyldithiocarbamates of the larger tetravalent actinide ions[241].

A diamagnetic, dimeric complex, $[NbCl(OEt_3).C_5H_6N]_2$, is obtained[242] by electrolytic reduction of the pentachloride in ethanol followed by the addition of pyridine. It reacts readily with sodium ethoxide to yield $Nb(OEt)_4$, which is also diamagnetic. Alcoholic pyridium chloride converts both compounds to the complex $(C_5H_6N)_2[Nb(OEt)Cl_5]$.

Pentavalent

Numerous pentavalent complexes are known and many of them are conveniently prepared by general reactions involving either the pentachlorides or pentaethoxides. Some typical examples of such reactions are illustrated in Fig. 8. These and other reactions are discussed in detail by Fairbrother[73] and only a few general comments are given here together with details of more recent studies.

Amides and Dialkyldithiocarbamates

The tantalum(V) pentakisdialkylamides ($Ta(NR_2)_5$ and $Ta(NRR')_5$) are more stable than their niobium(V) analogues. Thermal decomposition of the former, apart from $Ta(NMe_2)_5$ which is stable, leads to the formation[239] of pentavalent complexes of the type $RN{=}Ta(NR_2)_3$ (R = Et, ^{n}Pr and ^{n}Bu). Metal–nitrogen stretching vibrations for $Nb(NMe_2)_5$ and $Ta(NMe_2)_5$ are observed at 540 and 528 cm^{-1}, respectively (cf. 585 cm^{-1} for $Nb(NEt_2)_4$), whilst those for complexes of the type $RN{=}Ta(NR_2)_3$ range from 585 cm^{-1} (R=Et) to 634 cm^{-1} (R=^{n}Bu)[243]. In contrast to the behaviour of $Nb(NMe_2)_5$ discussed above, it is reported that the reaction between carbon disulphide and $Ta(NMe_2)_5$ yields the pentavalent complex $Ta(S_2CNMe_2)_5$ [241]. Infrared results suggest that the bonding in this compound is different from that in complexes for which it is known that both sulphur atoms of all the dialkyldithiocarbamates are coordinated and it will be interesting to have structural information on this complex.

Alkoxides and Related Complexes

In general the replacement of chlorine occurs more rapidly with alcohols than with primary or secondary amines. The thermal stability of the tantalum(V) penta-alkoxides is greater than that of the corresponding niobium(V) complexes and it is interesting to note that whereas stable tantalum(V) penta-tertiary alkoxides can be prepared by alcohol exchange reactions (Fig. 8), with niobium(V) products such as $NbO(O^tBu)_3$ are obtained[244]. However, $Nb(O^tBu)_5$ can be prepared by the alkoxide-tertiary butyl acetate exchange

239 D. C. Bradley and R. M. Thomas, *Can. J. Chem.* **40** (1962) 1355.
240 D. C. Bradley and M. H. Gitlitz, *J. Chem. Soc.* (A) (1969) 1152.
241 D. Brown, D. G. Holah and C. E. F. Rickard, *Chem. Commun.* 280 (1969); *J. Chem. Soc.* (A) 423 (1970).
242 R. A. D. Wentworth and C. H. Brubaker (Jr.), *Inorg. Chem.* **2** (1963) 551.
243 D. C. Bradley and M. H. Gitlitz, *J. Chem. Soc.* (A) (1969) 980.
244 D. C. Bradley, B. M. Chakravarti and W. Wardlaw, *J. Chem. Soc.* (1958) 4439.

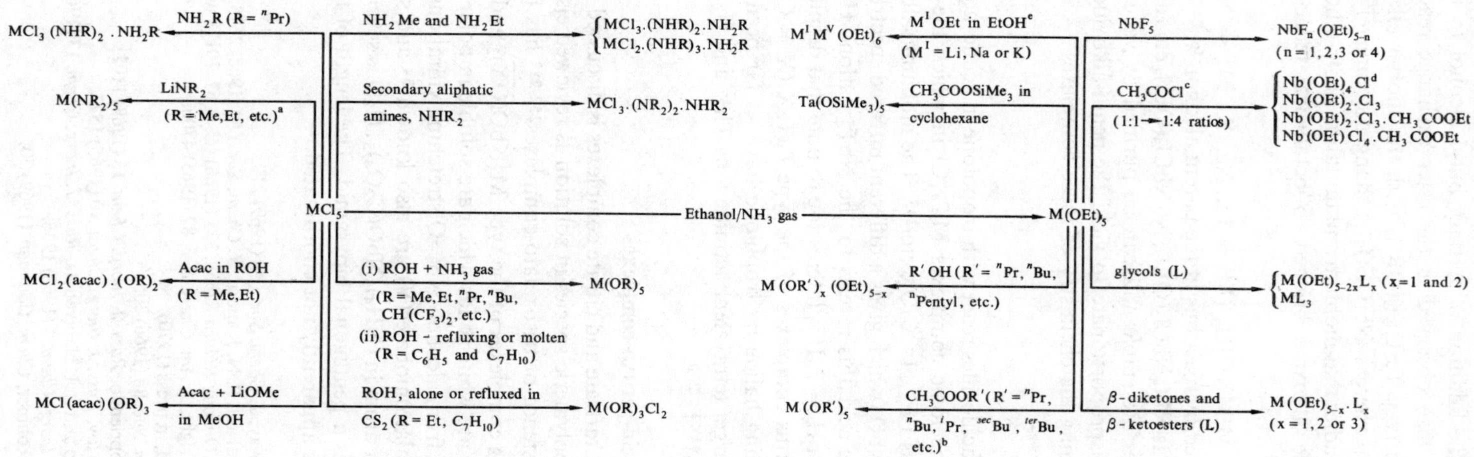

FIG. 8. Schematic representation of the preparation of some niobium and tantalum complexes.

[a] With longer chain amides reduction to niobium(IV) complexes occurs.
[b] With $Ta(OEt)_5$ mixed alkoxides are formed ($Ta(OEt)(OR')_4$) with all but the butylacetates.
[c] Similar reactions occur with acetyl bromide and also with other pentalkoxides.
[d] Also obtained by the reaction between $Nb(OEt)_5$ and dry HCl gas.
[e] This is a general reaction.

(Reproduced by permission of the copyright holder.)

reaction (Fig. 8) and also by the addition of tertiary butyl alcohol to niobium dialkylamides. Different behaviour is also observed in the ester exchange reactions, $Ta(OEt)_5$ yielding mixed alkoxides of the type $Ta(OEt)(OR')_4$ with the shorter alkyl chain acetates ($R' = {}^nPr$; ^iPr; ^nBu; ^iBu; ^nPn) whereas $Nb(OEt)_5$ exchanges completely. It has recently been shown by n.m.r. spectral studies that niobium and tantalum penta-ethoxides (R = Me, Et and ^{t}Bu) possess a double alkoxide bridged, bi-octahedral (edge-shared) dimeric structure[245].

Hexa-alkoxides of the type $M^IM^V(OR)_6$ (M^I = Li, Na, K; M^V = Nb and Ta; R = Me, Et, ^{i}Pr and ^{i}Bu) are prepared by reactions in the parent alcohol[246, 247] and some mixed chloro-alkoxide complexes such as $NEt_4NbCl_5(OEt)$, $NaNbCl_4(OEt)_2$ etc., are also known[247]. The tantalum(V) hexa-alkoxides are stable and can be purified by vacuum sublimation, whereas those of niobium(V) disproportionate to yield the penta-alkoxide ($Nb(OR)_5$) and M^IOR. Little is known concerning the chemistry of these complexes.

Tropolonates

Niobium and tantalum pentachlorides react with tropolone in acidic–aqueous or acidic–methanolic solution to form the ionic complexes $MT_4.Cl$ in which the chloride can be replaced by larger anions such as PF_6^-, I^-, ClO_4^-, etc.[248]. The niobium salts are hydrolysed in aqueous solution to yield T_3NbO (which gives a different infrared spectrum from T_3VO); on the addition of hydrochloric acid this reverts to the NbT_4^+ cation. However, the tantalum(V) chelate is stable up to pH = 7. It is interesting to note that despite its larger ionic radius protactinium(V) also forms complexes of the type PaT_4X (X = Cl, Br and I) and attempts to replace the remaining halide by tropolone have so far been unsuccessful[249]. In contrast to this the slightly larger tetravalent actinide ions Th^{4+} and U^{4+} (but not Np^{4+}) form complexes of the type $LiM^{IV}T_5$ [250].

α-Hydroxyacid and α-Hydroxyacid-peroxo-complexes

The earlier work on oxalate, tartrate and citrate complexes is discussed in detail by Fairbrother[73] and information on polymeric species in solution is reviewed elsewhere[251]. More recently the question of the existence of trisoxalato-complexes[252,253] has been resolved by the preparation[254] of anhydrous complexes of the type $M_3^INbO(\overline{OX})_3$ and $M_2^ITaOH(\overline{OX})_3$ (M^I = Ph_4P and Ph_4As). These are white solids which are soluble in polar organic solvents and which can be recrystallized from alcohol. Nb=O stretching vibrations are observed at 925 cm^{-1}. Hydrated oxytris(oxalato)niobates(V) are also known[254] and structural studies have shown[255] that the niobium atom in $(NH_4)_3NbO(C_2O_4)_3.H_2O$ is seven-coordinate, the oxygen atoms being arranged as a pentagonal bipyramid. The length of the Nb=O bond is 1·71 Å; the water molecule is apparently not coordinated.

245 D. C. Bradley and C. E. Holloway, *J. Chem. Soc.* (A) (1969) 219.
246 R. C. Mehrotra, M. M. Agrawal and P. N. Kapoor, *J. Chem. Soc.* (A) (1968) 2673.
247 L. Kolditz and M. Schoenterr, *Omagin Raluca Ripon*, **321** (1966) (*C.A.* **67**, 70178 w).
248 E. L. Muetterties and C. M. Wright, *J. Am. Chem. Soc.* **87** (1965) 4706.
249 D. Brown and C. E. F. Rickard, in press (1970).
250 E. L. Muetterties, *J. Am. Chem. Soc.* **88** (1966) 305.
251 A. N. Nevzorov and N. S. Onoprienko, *Russ. J. Inorg. Chem.* **13** (1968) 1102.
252 A. V. Vinogradov and I. V. Parlov, *Russ. J. Inorg. Chem.* **10** (1965) 1522.
253 A. K. Babko, E. A. Mazurenko and B. I. Nabivanets, *Russ. J. Inorg. Chem.* **13** (1968) 375.
254 A. Bričević and C. Djordjević, *Inorg. Chem.* **7** (1968) 1936.
255 G. Mathern, R. Weiss and R. Rohmer, *Chem. Commun.* (1969) 70.

Diperoxy-oxalato-complexes, $K_3M^V(O_2)_2(C_2O_4)_2 . H_2O$ (M^V = Nb and Ta) containing bidentate oxalate groups precipitate on the addition of oxalic acid to solutions of potassium niobate(V) or tantalate(V) in hydrogen peroxide[256]. Alternatively, the ammonium salt may be prepared by adding 30% hydrogen peroxide to $(NH_4)_3NbO(C_2O_4)_3 . H_2O$[255]. These complexes, like the triperoxy compounds $KM^V(O_2)_3$.dipy and $KM^V(O_2)_3$. *O*-phen.[256], are soluble in water and hydrogen peroxide but fail to dissolve in organic solvents. The two peroxide groups in $(NH_4)_3Nb(O_2)_2(C_2O_4)_2 . H_2O$ are in the *cis* configuration, the niobium atom being eight-coordinate with the oxygen atoms arranged as a distorted dodecahedron. The water molecule is apparently not coordinated to the niobium atom. Other peroxy complexes are discussed earlier (pages 588, 591 and 598).

Miscellaneous Complexes

Niobic acid (and to a lesser extent tantalic acid) reacts with a variety of aliphatic amines and with ethanolamines to form complexes containing aggregates of two to six units of niobic acid coordinated with one or two amine molecules. These complexes, and the earlier work on those isolated from solutions of niobic acid (or tantalic acid) in various higher α-hydroxy carboxylic acids and 1:2 diols are discussed in some detail by Fairbrother[73] and will not be dealt with here.

Few of the complexes used for the precipitation of niobium or tantalum during analytical procedures have been completely characterized but it is reported that the 8-quinolinol[257], cupferron[258, 259] and neocupferron[258] complexes are of the type M^VOL_3 (M^V = Nb and Ta; L = C_9H_6ON, $C_6H_5O_2N_2$ and $C_{10}H_7O_2N_2$).

2.17. ORGANOMETALLIC COMPOUNDS

The organometallic chemistry of niobium and tantalum is a relatively new field and fewer compounds are known than for vanadium. In addition, relatively few physical properties have yet been determined for the few compounds known and only a limited amount of structural information is available.

Cyclopentadienyl and Carbonyl Derivatives

Moisture-sensitive tetracyclopentadienyl compounds, $Nb(C_5H_5)_4$ and $Ta(C_5H_5)_4$, for which spectral evidence indicates that two cyclopentadienyl rings are π-bonded with two σ-bonded, are prepared by reduction of the appropriate pentachloride with a large excess of NaC_5H_5 in ethyl ether or benzene in a nitrogen atmosphere[260, 261]. Little is known concerning the physical properties of these compounds, which cannot be sublimed, apart from the fact that they are monomeric in benzene and that the dipole moment of $Ta(C_5H_5)_4$ is 4·09*D*. Anhydrous hydrogen chloride in ether converts the niobium compound to $(C_5H_5)_2NbCl_2$, a black volatile solid, which, according to Brantley[263], occurs as a by-product of the reaction between niobium pentachloride and NaC_5H_5 in dimethoxyethane.

256 C. Djordjević and M. Vuletić, *Inorg. Chem.* **7** (1968) 1864.
257 H. A. Szymanksi and J. H. Archibald, *J. Am. Chem. Soc.* **80** (1958) 1811.
258 Yu. I. Usatenko and A. P. Tikhonova, *Russ. J. Inorg. Chem.* **13** (1968) 797.
259 J. E. Land and J. R. Sanchez-Caldas, *J. Less-Common Metals*, **12** (1967) 41.
260 E. O. Fischer and A. Trieber, *Chem. Ber.* **94** (1961) 2193.
261 E. O. Fischer, *Angew. Chem.* (1960) 7235.
262 F. W. Siegert and H. J. De Liefole Meijer, *Rev. Trav. Chim.* **87** (1968) 1445.
263 J. C. Brantley, U. S. Pat. 2,921,948 (1960) (*C.A.* **54**, 11048).

The same author reports that when other inert solvents are used, this reaction yields $Nb(C_5H_5)_2$ and also briefly mentions compounds of the type $(C_5H_5)_2NbXOH$ (X = Cl and Br). On hydrolysis with aqueous hydrochloric acid $(C_5H_5)_2NbCl_2$ forms $(C_5H_5)_2Nb(OH)Cl_2$ which reacts with ammonium iodide in phenylmethanethiol to yield $(C_5H_5)_2NbI_2$ [264]. The mixed organometallate, $(C_5H_5)_2Nb(C_6H_5)_2$, a violet coloured, oxygen-sensitive solid, is formed on the addition of NaC_6H_5 to the dichlorodicyclopentadienide in benzene[262].

Pentavalent bis-π-dicyclopentadienyl trihalides of the type $(C_5H_5)_2M^VX_3$ (M^V = Nb and Ta; X = Cl and Br) are formed by the interaction of the appropriate pentahalide and NaC_5H_5, $Ca(C_5H_5)_2$ or $(C_5H_5)MgX_2$ (X = Cl or Br) in tetrahydrofuran or other solvents[263, 265, 266]. In the presence of an excess of $NaBH_4$ the white, air-sensitive tantalum trihydride $(C_5H_5)_2TaH_3$ is formed[267, 268] in high yield. N.m.r. spectral studies have indicated that two distinct types of Ta–H bond are present in this compound. The halide compounds are hydrolysed in aqueous solution with the formation of hydroxy derivatives, of general formula $(C_5H_5)_2M^VX_2OH$.

It has recently been shown that the reaction between $(C_5H_5)_2NbCl_2OH$ and hydrogen sulphide in methanol, in the presence of an ionic halide or pseudo-halide, results in the formation of a series of novel complexes containing the molecular unit S_2 as a ligand[269], $(C_5H_5)_2NbS_2X$ (X = Cl, Br, I and SCN). In addition to existing in a monomeric form the chloride and bromide are thought to possess polymeric modifications with disulphide bridging groups between niobium atoms. In the absence of an ionic halide the above reaction yields the polymeric sulphide $[(C_5H_5)_2NbS_4]_n$. On the basis of infrared spectral results it is suggested that canted π-cyclopentadienyl molecules are present in the polymeric compounds whereas parallel molecules are present in the polymeric chloride and bromide. The first example of a bis-π-cyclopentadienyl niobium(V) mixed halide, $(C_5H_5)_2NbI_2Cl$, has been prepared by the addition of methyl iodide to $(C_5H_5)_2NbS_2Cl$ in methylene dichloride. Small amounts of a white by-product, believed to be $(C_5H_5)_2NbOCl$, also form during this reaction.

Neither $Nb(CO)_5$ nor $Ta(CO)_5$ has been characterized, but salts containing the hexacarbonyl groups such as $M^IM^V(CO)_6$ [270] (M^I = alkali metal, M^V = V, Nb and Ta) and $[NaL_x][M^V(CO)_6]$[271, 272] (L = bis ethylene glycol dimethyl ether with $x = 2$, and L = dimethoxyethane with $x = 3$) have been prepared. The addition of organomercuric halides to sodium tris-dimethoxyethane tantalum hexacarbonyl (formal oxidation state −1) in benzene or ethyl ether solution results in the formation of diamagnetic compounds of the type $RHgTa(CO)_6$, where R = Me, Et or Ph, in which Ta–Hg bonds are believed to be present[273]. The stability of these derivatives decreases in the order Me > Et > Ph and only the first two can be sublimed unchanged in a vacuum.

Oxidation of the complexed sodium hexacarbonylmetallates (1−) with mercuric chloride in the presence of cyclopentadienylsodium converts them to the tetracarbonylcyclopentadienylmetallates (1+), $(C_5H_5)Nb(CO)_4$, m.p. 144–146°C, and $(C_5H_5)Ta(CO)_4$, m.p.

264 P. M. Teichel and G. P. Werber, *J. Organometallic Chem.* **12** (1968) 479.
265 G. Wilkinson and J. M. Birmingham, *J. Am. Chem. Soc.* **76** (1954) 4281.
266 J. Buloff, U.S. Pat. 3,030,393 (1962) (*C.A.* **57**, 7312).
267 J. A. McCleverty and G. Wilkinson, *Chem. and Ind.* (1961) 288.
268 M. L. H. Green, J. A. McCleverty, L. Pratt and G. Wilkinson, *J. Chem. Soc.* (1961) 4854.
269 P. M. Treichel and G. P. Werber, *J. Am. Chem. Soc.* **90** (1968) 1753.
270 J. E. Wymann, U.S. Pat. 3,232,699 (1966) (*C.A.* **64**, 12242h).
271 R. P. M. Werner and H. E. Podall, *Chem. and Ind.* (1961) 144.
272 R. P. M. Werner, A. H. Filbey and S. A. Manastyrskyj, *Inorg. Chem.* **3** (1964) 298.
273 K. A. Keblys and M. Dubeck, *Inorg. Chem.* **3** (1964) 1646.

171–173°C, respectively[272]. Small yields of the niobium compound have also been obtained by heating together the pentachloride and cyclopentadienylsodium in an atmosphere of carbon monoxide in the presence of hydridic reducing agents. With $LiBH_4$ as the reducing agent in the presence of hydrogen and carbon monoxide $(C_5H_5)_2NbClBH_4$ was obtained[274]. The dark red, air-sensitive tetracarbonylcyclopentadienylmetallates (1+) react with triphenyl phosphine, arsine and stilbene under the influence of ultraviolet radiation to form complexes of the type $(C_5H_5)NbPh_3X$[275] (X = P, As or Sb) and with diphenylacetylene (Tolan = T) to form compounds such as $(C_5H_5)Nb(CO)_2T$[276], $(C_5H_5)M^V(CO)T_2$[277, 278] (M^V = Nb and Ta) and $(C_5H_5)Nb(CO)T_3$[277]. When $(C_5H_5)Nb(CO)_2T$ is heated in toluene in the absence of ultraviolet radiation the dimer $[(C_5H_5)Nb(CO)T]_2$ is obtained. Single-crystal studies have shown that $(C_5H_5)Nb(CO)T_2$ and $[(C_5H_5)Nb(CO)T]_2$ both possess triclinic symmetry (Table 28). Two independently coordinated tolan molecules are present

TABLE 28. CRYSTALLOGRAPHIC DATA FOR SOME ORGANOMETALLIC COMPOUNDS OF NIOBIUM

Compound*	Symmetry	Space group	Lattice parameters						Reference
			a_0 (Å)	b_0 (Å)	c_0 (Å)	$\alpha°$	$\beta°$	$\gamma°$	
$(C_5H_5)Nb(CO)T_2$	Monoclinic	$P\bar{1}$	11·72	11·80	10·06	100·5	103·4	78·6	a
$[(C_5H_5)Nb(CO)T]_2$	Triclinic	$P\bar{1}$	10·43	9·60	9·12	110·3	71·1	114·1	b

* T = diphenylacetylene (tolan).
[a] A. N. Nesmeyanov, A. I. Gusev, A. A. Pasynskii, K. N. Anisimov, N. E. Kolobova and Yu. T. Struchkov, *Chem. Commun.* (1969) 277.
[b] A. N. Nesmeyanov, A. I. Gusev, A. A. Pasynskii, K. N. Anisimov, N. E. Kolobova and Yu. T. Struchkov, *Chem. Commun.* (1968) 1365.

in the former, Nb–C bond distances range from 2·17 to 2·23 Å, and the mean Nb–C bond distance for the cyclopentadienyl molecule is 2·46 Å[279]. Bridging tolan molecules are present in $[(C_5H_5)Nb(CO)T]_2$ which provides the first example of this type of coordination and, in addition, the first example of an organometallic compound containing niobium–niobium bonds[280]. The Nb–Nb bond distance, 2·74 Å, is less than those in, for example, $NbCl_4$ (3·06 Å), NbI_4 (3·31 Å) and the various halogen cluster compounds described earlier (Tables 10 and 12). The structure of this compound is illustrated in Fig. 9. Structural data are also available for $(C_5H_5)Nb(CO)_4$ [281].

[274] R. B. King, *Z. Naturforsch.* **18B** (1963) 157.
[275] A. N. Nesmeyanov, K. N. Anisimov, N. E. Kolobova and A. A. Pasynskii, *Izv. Akad. Nauk. SSSR, Ser. Khim.* (1966) 2231.
[276] A. N. Nesmeyanov, K. N. Anisimov, N. E. Kolobova and A. A. Pasynskii, *Izv. Akad. Nauk. SSSR, Otdel Khim. Nauk.* (1966) 774.
[277] A. N. Nesmeyanov, K. N. Anisimov, N. E. Kolobova and A. A. Pasynskii, *Izv. Akad. Nauk. SSSR, Otdel Khim. Nauk.* (1969) 100.
[278] A. A. Pasynskii, K. N. Anisimov, N. E. Kolobova and A. N. Nesmeyanov, *Izv. Akad. Nauk. SSSR, Otdel Khim. Nauk.* (1969) 183.
[279] A. N. Nesmeyanov, A. I. Gusev, A. A. Pasynskii, K. N. Anisimov, N. E. Kolobova and Yu. T. Struchkov, *Chem. Commun.* (1969) 277.
[280] A. N. Nesmeyanov, A. I. Gusev, A. A. Pasynskii, K. N. Anisimov, N. E. Kolobova and Yu. T. Struchkov, *Chem. Commun.* (1968) 1365.
[281] H. W. Baird, Thesis, Wisconsin Univ. (1963), according to R. J. Doedens and L. F. Dahl, *J. Am. Chem. Soc.* **87** (1965) 2576.

Miscellaneous Compounds

The trimethyldichloro-compounds Me_3NbCl_2 and Me_3TaCl_2 are obtained by low-temperature reactions involving dimethyl zinc and the appropriate pentachloride[282]. They are air- and moisture-sensitive solids; Me_3TaCl_2 is thermally less stable than its niobium analogue which sublimes readily *in vacuo* at room temperature to form golden-yellow crystals. Evidence for the existence of C–Nb bonds has been obtained by proton n.m.r. studies on solutions of the compound in carbon tetrachloride.

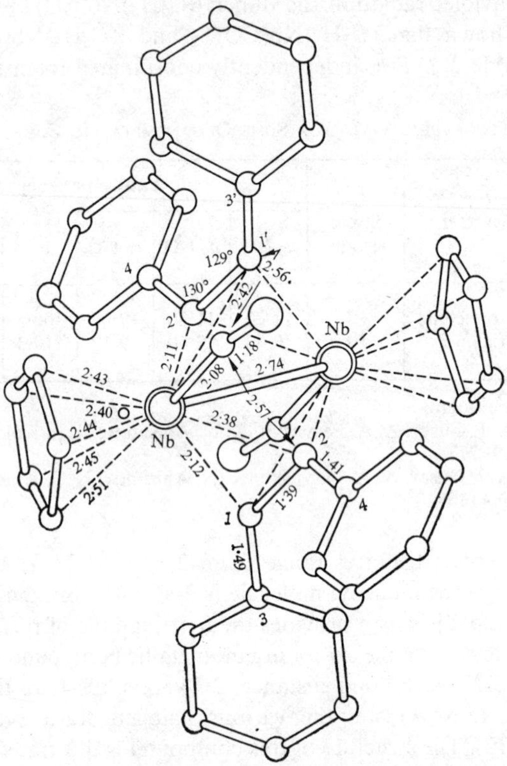

FIG. 9. The structure of $[(C_2H_5)Nb(CO)T]_2$ [280].
(Reproduced by permission of the copyright holder.)

Hexamethylbenzene chloro-compounds $Nb_2(C_6Me_6)_2Cl_4$, $[Nb_3(C_6Me_6)_3Cl_6]Cl$ and $[Ta_3(C_6Me_6)_3Cl_6]Cl$ have been prepared by interaction of hexamethylbenzene and the appropriate pentachloride in the presence of aluminium and aluminium trichloride[283] and the existence of difluorenylniobium trichloride, $(C_{13}H_9)_2NbCl_3$, has been briefly reported[284]. The latter is converted to $(C_{13}H_9)NbCl_3Br$ on treatment with bromine in carbon tetrachloride.

[282] G. L. Junivall, *J. Am. Chem. Soc.* **86** (1964) 4202.
[283] E. O. Fischer and F. Roehrscheid, *J. Organometallic Chem.* **6** (1966) 53.
[284] R. D. Gorisch, U.S. Pat. 3,080,305 (1963) (*C.A.* **59**, 3957).

INDEX

Contents of Comprehensive Inorganic Chemistry

VOLUME 1

The element Hydrogen, ortho- and para-Hydrogen, atomic Hydrogen
K M MACKAY *University of Waikato*

Hydrides
K M MACKAY *University of Waikato*

Deuterium and Tritium
K M MACKAY *University of Waikato* & M F A DOVE *University of Nottingham*

Proton, protonic acids and hydrogen bond
J E PRUE *University of Reading*

The monatomic gases: physical properties and production
A H COCKETT & K C SMITH *British Oxygen Co Ltd, London*

The chemistry of Krypton, Xenon and Radon
N BARTLETT *University of California, Berkeley* & F O SLADKY *University of Innsbruck*

Lithium and its compounds
W A HART & O F BEUMEL Jr *Foote Mineral Co, Pennsylvania*

Sodium, Potassium, Rubidium, Cesium and Francium
T P WHALEY *International Minerals and Chemical Corp, Illinois*

Beryllium
D A EVEREST *National Physical Laboratory, Teddington*

Magnesium, Calcium, Strontium, Barium and Radium
R D GOODENOUGH & V A STENGER *The Dow Chemical Co, Michigan*

Boron
N N GREENWOOD *University of Leeds*

Aluminium, Gallium, Indium and Thallium
K WADE & A J BANISTER *University of Durham*

Carbon
A K HOLLIDAY, G HUGHES & S M WALKER *University of Liverpool*

An introduction to the organic chemistry of the metallic elements
M L H GREEN *University of Oxford* & P POWELL *Royal Holloway College*

Silicon
E G ROCHOW *Harvard University*

Index

VOLUME 2

Germanium
E G ROCHOW, *Harvard University*

Tin
E W ABEL *University of Exeter*

Lead
E W ABEL *University of Exeter*

Nitrogen
K JONES *University of Manchester Inst. of Science & Technology*

Phosphorus
A D F TOY *Stauffer Chemical Co, New York*

Arsenic, Antimony and Bismuth
J D SMITH *University of Sussex*

Oxygen
E A V EBSWORTH *University of Edinburgh*,
J A CONNOR *University of Manchester*, &
J J TURNER *Newcastle on Tyne University*

Sulphur
M SCHMIDT & W SIEBERT *University of Wurzburg*

Selenium, Tellurium and Polonium
K W BAGNALL *University of Manchester*

Fluorine
T A O'DONNELL *University of Melbourne*

Chlorine, Bromine, Iodine and Astatine
A J DOWNS & C J ADAMS *University of Oxford*

Index

VOLUME 3

Copper
A G MASSEY *University of Technology, Loughborough*

Silver
N R THOMPSON *ICI, Runcorn Heath Laboratory, Cheshire*

Gold
B F G JOHNSON *University of Cambridge*
& R DAVIS *Kingston Polytechnic*

Group IIB
B J AYLETT *Westfield College, London*

Scandium, Yttrium and Lanthanum
R C VICKERY *Hudson, Florida*

Titanium
R J H CLARK *University College, London*

Zirconium and Hafnium
D C BRADLEY & P THORNTON *Queen Mary College, London*

Vanadium
R J H CLARK, *University College, London*

The chemistry of Niobium and Tantalum
D BROWN *Atomic Energy Research Establishment, Harwell*

Chromium, Molybdenum and Tungsten
C L ROLLINSON *University of Maryland*

Manganese
R D W KEMMITT *University of Leicester*

Technetium
R D PEACOCK *University of Leicester*

Rhenium
R D PEACOCK *University of Leicester*

Iron
D NICHOLLS *University of Liverpool*

Cobalt
D NICHOLLS *University of Liverpool*

Nickel
D NICHOLLS *University of Liverpool*

The 2nd and 3rd row elements of group VIIIA, B and C
S E LIVINGSTONE *University of New South Wales*

Index

VOLUME 4

The Lanthanides
T MOELLER *Arizona State University*

Carbonyls, Cyanides, Isocyanides and Nitrosyls
W P GRIFFITH *Imperial College of Science & Technology London*

Compounds of the transition elements involving metal-metal bonds
D L KEPERT *University of Western Australia*
& K VRIEZE *University of Amsterdam*

Transition metal hydrogen compounds
J C GREEN & M L H GREEN *University of Oxford*

Non-stoichiometric compounds: an introductory essay
D J M BEVAN *The Flinders University of South Australia*

Tungsten bronzes, Vanadium bronzes and related compounds
P HAGENMULLER *University of Bordeaux*

Isopolyanions and heteropolyanions
D L KEPERT *University of Western Australia*

Transition metal chemistry
B F G JOHNSON *University of Cambridge*

Organo-transition metal compounds and related aspects of homogeneous catalysis
B L SHAW & N I TUCKER *University of Leeds*

Index

VOLUME 5

The Actinides
J A LEE & J A MARPLES
G W C MILNER & G PHILLIPS
P G MARDON
Atomic Energy Research Establishment, Harwell
C KELLER *Kernforschungszentrum, Karlsruhe*
S H EBERLE *Universität Karlsruhe*
D BROWN, R M DELL & P E POTTER *AERE, Harwell*
K W BAGNALL *University of Manchester*
J RYDBERG & J O LILJENZIN *Chalmers University of Technology*
S AHRLAND *University of Lund*

Master Index

INDEPENDENT OPINION

"These books are attractively bound and have clear print. Since the length and cost are not prohibitive, this set of books should be well within the budget of most libraries. Not only will the professional chemist find these books useful, but students and other readers will find them a valuable reference source. (Comprehensive Inorganic Chemistry) should be found in every undergraduate and graduate library, as well as industrial libraries. Many professional chemists may even consider them for personal libraries. Highly recommended."

Choice—*A publication of the Association of College and Research Libraries.*

INDEPENDENT OPINION

Volume 1 1467 pp + index

"This covers the chemistry of hydrogen, the noble gases, and of the elements of Groups IA, IIA, IIIB, carbon and silicon. The first three chapters deal with hydrogen, hydrides, deuterium and tritium and the fourth is an interesting discussion of the proton, protonic acids and the hydrogen bond. Two chapters follow on the inert gases, including interesting and extensive recent knowledge about their compounds set out by N. Bartlett and F. O. Sladky. Four chapters on the alkalis and alkaline earths contain a wealth of detail, although perhaps along traditional lines. N. N. Greenwood has written an excellent account on boron chemistry of book length in itself, and another chapter deals at length with much new informationabout aluminium, gallium, indium and thallium. Chapters 13 and 15 deal with carbon and silicon. Here it might have been expected that more would have been included on the high pressure chemistry of carbon and silicates, and mineral chemistry. Chapter 14 by M. L. H. Green and P. Powell is a useful introduction to the organic chemistry of the metallic elements, along modern lines of ligand field theory and ideas about metal complexes.

Throughout this volume, with its different authors, it is perhaps inevitable that there is some lack of uniformity in the extent of detail given. There are also a few lapses in symbolism, notation, and uniformity of units. Yet the whole must be regarded as a highly commendable collection of material which will be valuable to chemists of all kinds."

Professor Sir Harold Thompson FRS
Oxford

Volume 2 1594 pp + index

"Volume 2 is concerned with the chemistry of the elements of Groups IV, V, VI, VII. The general impression on reading the various chapters of this volume is the great effectiveness in reporting a considerable amount of chemistry in a very digestible form. The systematic presentation applied to each chapter allows a rapid assessment of the appropriate chemical information, and the text is well documented with reference to the original literature plus good review articles for a more detailed coverage. Perhaps a minor criticism of this work lies in the indexing; the subject index is relatively sparse for a text of such magnitude and it would have been of considerable utility to have a formulae index to the text.

The area of chemistry covered by this work is obviously very great, but it does appear to have dealt with it in a very succinct manner for the majority of the text, which extends to approximately 1500 pages. The two major chapters in the book are associated with the chemistry of nitrogen, approximately 240 pages, and the chemistry of the halogens (excepting fluorine), approximately 500 pages. The book thus encompasses in one volume what would normally be a series of books. Both of the above mentioned chapters are admirable and the authors, K. Jones on nitrogen with A. J. Downs and C. J. Adams on the halogens, are to be commended on both the presentation and coverage. The chapter on the halogens illustrates the real strength of the series, in that detailed chemical information is not only presented but discussed in physico-chemical theoretical terms. A scientific compendium of this size often suffers from the "catalogue" approach, but the present text presents the chemistry in critical mode with a realistic assessment of the various physical methods used in property determination. Thus the properties of the halogens are discussed in terms of bond energies, bond lengths, vibrational properties, e.s.r., n.m.r., n.q.r. and Mossbauer spectroscopy, electronic and magnetic properties and dipole moments allowing a detailed appraisal of the use of various modern methods in studying the chemical properties reviewed.

Considering the magnitude of the task undertaken, it is extremely pleasing to note the number of chapters referring to papers in the 1971 period—a truly great commendation on the overall editorship of these volumes. Perhaps a general note in each volume stating the period covered by the references would have been of help. In general this work provides a welcome and unique addition to the inorganic literature."

Professor J. Lewis FRS
Cambridge

Volume 3 1370 pp + index

"This volume covers the chemistry of the elements of the d-block of the Periodic Table (the transition elements), with the exception of the Lanthanide elements (Vol. 4), the Actinide elements (Vol. 5) and some special aspects which are common to many of the transition elements (Vol. 4). The volume is therefore concerned specifically with the three elements which characterize each of the ten transition groups, and the chapters are mostly grouped in this way. However, the six platinum metals are treated in one chapter which is the best way to fit these similar elements into the overall scheme which is standard for all five volumes. There are altogether 17 chapters, written by 14 authors who are internationally recognized 1370 pages of text and a useful 17 page index.

The five volumes are quite remarkable, in that they can justifiably claim to be comprehensive, yet at the same time remain interesting and readable; they are probably unique in this respect. Volume 3 serves as an excellent source-book for the essential physical constants of all important compounds (simple and complex) of the transition metals. These are arranged so that significant comparisons are made wherever possible, and there are extensive references. It says much for the ingenuity of the editors, authors, and particularly the printers that the presentation of such an amount of information has been possible, while still maintaining the readability of the text. Throughout the volume chemical properties and reactions are discussed and interpreted rather than listed. The need for skilled correlation of data is particularly important in Volume 3, since it is in the area of the transition elements that a major part of the research work in inorganic chemistry has been published in recent years, and in this area also there has been a major interaction of inorganic with theoretical chemistry.

This volume must surely become the first point of reference for research workers and teachers alike. The transition elements play an important role in Pure and Applied Chemistry, Physics, Materials Science and Biology, and the authors clearly intend their chapters to be of value to this wide audience. Teachers at any level will also appreciate the very high quality of the general presentation, discussion, formulae and diagrams. Apart from reference to the original literature, few scientists will find it necessary to look outside this volume for their material."

Professor C. C. Addison FRS
Nottingham

Volume 4 994 pp + index

"Volume 4 is concerned with the general chemistry of the lanthanides and some special topics in transition metal chemistry.

Therald Moeller has packed a great amount of the fundamental chemistry of the lanthanides into his 101 pages in an interesting and scholarly manner with tables of essential data. Important recent developments in their organometallic chemistry have come too late to be included, but the chapter provides a useful fairly detailed first reference to their inorganic chemistry.

The subjects of the surveys are topical and obviously bear the mark of the late Sir Ronald Nyholm. They vary considerably in detail of treatment, interest and authority. Generally they emphasize recent work until about 1969–70 but rarely show a sense of history. They vary in length from 60 to 200 pages, mostly around 100 pages. They are authoritative and useful surveys all giving numerous references to recent reviews and original work. The authors are well known chemists whose style and subject matter are familiar to most inorganic chemists. There are eight surveys as follows:—

Carbonyls, cyanides, isocyanides and nitrosyls by W. Griffith. Compounds of the transition elements involving metal-metal bonds by D. L. Kepert and K. Vrieze. Transition metal hydrogen compounds by J. C. Green and M. L. H. Green. Non-stoichiometric compounds: an introductory essay by D. J. M. Bevan. Tungsten bronzes, vanadium bronzes and related compounds by P. Hagenmuller. Isopolyanions and heteropolyanions by D. L. Kepert. Transition metal chemistry by B. F. G. Johnson. Organo-transition metal compounds and related aspects of homogeneous catalysis by B. L. Shaw and N. I. Tucker.

This volume has its own subject index of sixteen and a half pages, and is well produced with numerous tables of data and references provided at the foot of each page."

Professor J. Chatt FRS
Sussex

Volume 5 635 pp + Master index

"Volume 5 is devoted to the Actinides (635 pp) and the Master Index (78 pp). The latter serves little purpose since it merely indicates the subsections of CIC, and thus repeats the indexes in each individual volume. Indeed, as the treatment of each element or series of elements follows a standard pattern, the volumes are essentially self-indexing anyway. A one-page table of contents at the beginning of Volume 5 would have been more helpful and is a curious omission. The running headings at the top of each double page are also singularly uninformative, only three being used: 'The Elements' for 102 pages, 'Compounds' for 361 pages and 'Solution Chemistry' for the remaining 171 pages.

The treatment of actinium and the actinides (elements 89–103) is both readable and authoritative. Nine of the contributors are from AERE, Harwell, and the other five (with one exception) are from nuclear chemistry institutes in Sweden and Germany. In reviewing these 5f elements it is salutary to recall that the majority have been synthesized for the first time within the last 30 years—yet the number of compounds known and the amount of information on them has already outstripped the more limited chemistry of their 4f congeners, the lanthanides. The authors have done a magnificent job in assembling, collating, assessing, and systematizing a vast amount of data on the physical and chemical properties of these elements and their numerous compounds. The work, which is extensively referenced, will undoubtedly remain the standard first source of information in this area for many years to come."

Professor N. N. Greenwood FRIC
Leeds

Periodic Tabl

Transition Elen

Group:		IA	IIA	IIIB	IVB	VB	VIB	VIIB
Principal quantum number (Period)	Valence shell	s^1	s^2	$d^1s^2f^x$	d^2s^2	$(d^3s^2)^†$	$(d^5s^1)^†$	d^5s^2
$n = 1$	$1s$	1 H 1.008						
$n = 2$	$2s2p$	3 Li 6.94	4 Be 9.01					
$n = 3$	$3s3p$	11 Na 22.99	12 Mg 24.31					
$n = 4$	$4s3d4p$	19 K 39.10	20 Ca 40.08	21 Sc 44.96	22 Ti 47.90	23 V 50.94	24 Cr 51.996	25 Mn 54.94
$n = 5$	$5s4d5p$	37 Rb 85.47	38 Sr 87.62	39 Y 88.91	40 Zr 91.22	41 Nb 92.91	42 Mo 95.94	43 Tc 99
$n = 6$	$6s4f5d6p$	55 Cs 132.91	56 Ba 137.34	57* La 138.91	72 Hf 178.49	73 Ta 180.95	74 W 183.85	75 Re 186.2
$n = 7$	$7s5f6d7p$	87 Fr	88 Ra 226	89⊙ Ac				

*Lanthanide Series	58 Ce 140.12	59 Pr 140.91	60 Nd 144.24
⊙Actinide Series	90 Th 232.04	91 Pa	92 U 238.03

†Variable valence shells